（英）小弗（Frances Arnold）著

辽宁科学技术出版社
·沈阳·

图书在版编目（CIP）数据

上海玻璃博物馆 / （英）小弗（Frances, A.）著.
沈阳：辽宁科学技术出版社，2014.1
　　ISBN 978-7-5381-8430-3

　　I. ①上… II. ①小… III. ①玻璃—博物馆—建筑设
计—上海市　IV. ①TU242.5

　　中国版本图书馆CIP数据核字(2013)第312278号

总　策　划：《城市·环境·设计》杂志社 (UED Magazine)
撰　　　稿：小弗 (Frances Arnold)
设计指导：昕羽 (Jessica Hoecke)
前言作者：彭礼孝、简必珂 (Clare Jacobson)
特别鸣谢：协调（亚洲）(COORDINATION ASIA Ltd)
　　　　　罗昂建筑设计 (logon urban. architecture. design)
　　　　　上海玻璃博物馆 (Shanghai Museum of Glass)
封面摄影：Die Photodesigner 摄影 (Die Photodesigner)

出版发行：辽宁科学技术出版社
　　　　　（地址：沈阳市和平区十一纬路29号　邮编：110003）
印　刷　者：北京雅昌彩色印刷有限公司
经　销　者：各地新华书店
幅面尺寸：280 mm × 280 mm
印　　张：14.5
插　　页：4
字　　数：30 千字
出版时间：2014 年 1 月第 1 版
印刷时间：2014 年 1 月第 1 次印刷
责任编辑：付　蓉
文字编辑：王贝贝
封面设计：李洁璇
版式设计：李洁璇
责任校对：王玉宝

书　　号：ISBN 978-7-5381-8430-3
定　　价：298.00元

序言

博物馆是城市的眼睛，它以丰富的知识和深厚的积淀，折射出一座城市的灵魂和精神。

作为公众获取知识、提高修养的精神殿堂，博物馆在中国的作用日渐突出，从边缘步入常态，在人们的日常生活中占据着越来越重要的地位。尤其是近年来随着人民物质水平和审美意识的提高，中国的博物馆建筑、美术馆建筑和展览馆建筑的设计建造有着突飞猛进的增长，博物馆的新职能、新形态、新方法、新主题也不断地涌现出来。国内外建筑师针对中国的现实需求，进行了大量的研究及实践，设计出了一批优秀的博物馆建筑。但是，从某种程度上看，我国的博物馆设计仍然存在着诸多不容忽视的问题，如建筑与展陈的脱离、建筑形体与当地文化的冲突、建筑师设计概念与甲方想法相左等。

《城市·环境·设计》（以下简称UED）杂志社于2011年设立的"UED博物馆建筑设计奖"是面向中国博物馆建筑设计界的权威奖项，以鼓励在博物馆建筑设计与城市、展陈、文化的关系等多个方面进行深入研究及实践的中国建筑师。UED与上海玻璃博物馆的结缘正是在"2013年度UED博物馆建筑设计奖"的评审会上，上海玻璃博物馆不负众望，顺利入围。从建筑学角度来讲，该作品颠覆了传统意义上的博物馆概念，极大地丰富了博物馆的内涵与外延，而在社会层面，该博物馆促使当代博物馆在管理属性、经营策略、发展目标等方面有了不同程度的转变，因此获得了建筑界对其充分的肯定。

本书以主题性博物馆建筑作为研究视角，以大量篇幅论述玻璃博物馆之于传统博物馆的不同之处，围绕着"分享玻璃的无限可能"的议题，对于玻璃博物馆从城市规划、博物馆建筑、博物馆策划、博物馆设计、博物馆品牌等方面进行详尽的阐述与总结，彰显其在新时代背景下自身所拥有的设计价值、艺术价值以及社会价值等综合性意义。这主要体现在以下几个方面：第一，本书从城市规划和博物馆建筑的角度来展示对上海轻工玻璃厂的宝山区工厂旧址进行改造的社会意义，即建筑在完成了连接历史、传统、现代、未来的使命的同时，也体现了后工业时代对于工业遗产的再利用以及其在艺术区与工业区之间的成功转型；第二，本书从博物馆策划和博物馆设计的角度来阐述玻璃博物馆脱离了单纯的建筑学本体性意义而延伸至建筑背后的文化内涵与社会价值，即伴随着经济的发展与时代的进步，人类的精神文明和审美水平也逐渐提高，博物馆由原来仅仅是一个收集、贮存和处理信息的容器转变成普及知识、提高修养、陶冶情操的精神场所；第三，上海玻璃博物馆不同于一般性质的博物馆，它采用的不是一种让公众被动接受信息的参观方式，而是强调互动性和体验性，通过参观流线的趣味性、参观模式的多样性和参与内容的丰富性，对传统空间设计观进行一定的调整与完善，使博物馆既能表现多元化的审美趣味，又能促成人与人之间的沟通与交流；第四，从品牌建设的角度看，上海玻璃博物馆重点诠释了博物馆"在工业环境中培育文化热点"的品牌战略以及"以人为本"的服务理念，全面而系统地展示了当代博物馆建设的策略性与整体性特征。

总的来说，上海玻璃博物馆以合理的体量、适宜的尺度、辨识性强的外观和内涵丰富的社会定位延展了博物馆建筑的基本功能与内在价值，是值得肯定和传播的。

彭礼孝
《城市·环境·设计》(UED)杂志社主编
天津大学建筑学院特聘教授
2013年10月于北京

PREFACE

A museum forms the eyes of a city. It shows the reflection of its soul and spirit, with a rich accumulation of knowledge and tradition.

As a spiritual palace for the public to absorb knowledge and to improve self-cultivation, museums play an increasingly important role in China and have become integrated into people's day-to-day existence. Recent years have seen improvements in both quality of life and in aesthetic awareness, resulting in rapid growth within the fields of museums, galleries and exhibitions, seeing them embrace new functions, forms, approaches and themes.

Increasingly, domestic and foreign architects are responding to this demand by creating iconic buildings to house these new institutions. However, problems continue to arise: the separation between building and exhibition, for example; conflicts between the architectural form and local culture; controversy between the architects' design concept and the clients' ideas, and so on.

Urban • Environment • Design (hereinafter referred to as UED) Magazine established the "UED Museum Building Design Award" in 2011. A prestigious architectural prize, it seeks to encourage architects committed to exploring the relationships between museum design, cities, exhibitions and culture. The Shanghai Museum of Glass was shortlisted for the 2013 edition of "UED Museum Architectural Design Award", earning the project industry-wide recognition. Architecturally speaking, the work breaks away from the traditional concept of museums, greatly enriching the connotations and extension for venues of this kind. More than that, the project presents a new model for contemporary museums in terms of management attributes, business strategies and development goals.

In this book, through the study of the thematic museum architecture, the differences between the Shanghai Museum of Glass and traditional museums are discussed in detail. Focusing on the "countless possibilities of glass", it is a detailed description and summary of urban planning and museum architecture, planning, design and branding, highlighting the design, artistic and social values of the venue in the context of a new era.

The work explores the significance of transforming the former premises of Shanghai Glass Company Ltd. in terms of urban planning and museum architecture. The result references the historical, traditional, contemporary and future missions of the building, all the while presenting a prime example of the reutilization of industrial heritage, and the successful transformation from factory area to art district in the post-industrial era. Beyond its architectural links, the museum effectively extends the cultural connotations and social values of Shanghai Glass Company Ltd. both through economic development as well as on a more spiritual level. This forms part of an ongoing trend which sees museums gradually shift from mere containers for information, storage and processing, into a living place for learning and self-cultivation. To that end, the Shanghai Museum of Glass differs from ordinary museums through its emphasis on interactivity and experience, as opposed to passive acceptance of information. With interesting visitor routes, diversified displays and rich content, the venue improves upon traditional spatial design concepts to encompass a broad range of aesthetic tastes, as well as promote communication and exchange. The museum's innovation extends to its branding, cultivating a forum for dialogue in an industrial setting, all the while maintaining a highly people-oriented service concept.

In summary, both as a cultural institution and case study for development, the Shanghai Museum of Glass is worthy of our attention. Combining a breadth of content, iconic architecture and community focus, the museum builds on the basic functionality and intrinsic value of museums, setting a precedent for the future.

Peng Li Xiao
Editor-in-chief, UED
Distinguished Professor of School of
Architecture, Tianjin University

前言

如今正在中国兴起的博物馆建筑热潮已经成为各类媒体的热门话题,这通常不是一件好事。美国的国家公共广播电台报道:"中国建造了众多博物馆,但展示什么内容却又是另外一回事。"[1] 英国《卫报》报道:"中国博物馆藏品中多达40 000件赝品。"[2] 中国的《人民日报》指出:"免费参观也无法吸引人们踏入博物馆。"[3] 不过,在中国还有另外一类有关新建的、成功的博物馆的新闻,通常登不上头版,只能安静地作为那些头条新闻的陪衬,但是我们能从这些博物馆的身影中看到中国博物馆兴建热潮的正确方向。

上海玻璃博物馆便是此类博物馆中的一员。与其他声名欠佳的博物馆所不同的是,它将建筑与展示内容共同考虑,对参观者与周边邻里一视同仁,将博物馆的第一印象与长期发展等量齐观,正是这些基本理念成就了博物馆的一举成功。而这样的成功并非前所未有,相反,它正是在汲取了中国其他新建博物馆经验教训的基础上而取得的成功。遵循简单而又核心的原则,上海玻璃博物馆虽然没有成为新闻热点,但却是最贴近当今中国现实的博物馆。

首先,这个博物馆绝非仅仅意味着展览的空间。用"购置、维护、研究和展示具有持久趣味和价值的物品的机构"来定义博物馆,这在21世纪似乎已经有点过时。[4] 在墙面上布满画作、展示丰富藏品诚然是任何一家当代博物馆的核心,不过如今的博物馆内涵尚需拓展,从而在文化和经济层面上得以维系。从这一点上来说,中国的博物馆所面临的现实与世界上其

他的博物馆都一样——来自其他文化场馆的竞争加剧,而博物馆传统的资金来源却在减少。上海一些新的博物馆意识到了上述问题,进而扩充了内容策划和资金来源:外滩美术馆举办一系列晚间讲座和表演;上海当代艺术博物馆为一些赞助活动提供场地;上海自然博物馆还合并了一个中央公园。上海玻璃博物馆同样需要超越单纯的藏品展示的传统手段。与上海的其他博物馆不同,上海玻璃博物馆远离市中心,也不属于传统的旅游观光线路,远道而来的参观者非常期待他们的参观之旅物有所值。为了满足这样的诉求,博物馆采取了互动展示的方式,而不是令观众单纯观赏艺术品。除了主展览厅,还有为全天游览行程而特别策划的辅助空间,包括热玻璃演示、博物馆咖啡厅和DIY创意工坊。此外,各类活动空间为博物馆开拓了那些博物馆常客之外的受众,同时也带来了额外的收益。

其次,上海玻璃博物馆成功的另一个关键是对工业建筑的再利用。在全球范围内很多新博物馆非常流行采用此类建筑改建,因为其坚固的建筑外壳能为当代艺术提供开阔的空间和充足的自然光。而且,在老旧工业建筑躯体上开发新的博物馆,将博物馆与这个地区过往的历史构建联系,这本身就是一种可持续发展的模式。与此同时,与拆除现场老结构、重新设计施工相比,这种方案也更为经济。伦敦的泰特现代美术馆(Tate Modern,原为发电厂)和美国北亚当斯的马萨诸塞州当代艺术博物馆(Massachusetts Museum of Contemporary Art,原为印染厂)都利用了工业建筑进行改建,在国际上

享有盛誉。而毗邻的案例有上海当代艺术博物馆(改建自旧电厂)和民生现代美术馆(原为钢铁厂)。上海玻璃博物馆前身是上海轻工玻璃厂的生产总部,是博物馆反映老建筑原有行业形态的经典案例。这种新旧联系从某种程度上说是强制性的,因为在大部分玻璃生产车间都已迁出原址之后,中国的土地法规要求场地仍旧保持与玻璃产业相关的用途。事实上,如今与旧址最直接的联系就是老厂房建筑的保留,而这种新与旧的关联也恰到好处地在博物馆的设计和策划中体现了出来。

再者,上海玻璃博物馆的成功在于它并不仅仅想要成为一个地标性建筑。虽然博物馆的设计旨在超越这个区域,进而成为宝山区的中心,但设计更想和谐地融入周边区域,而非标新立异地塑造一个具有巨大反差的建筑。标志性的建筑并非意味着博物馆项目的失败,像纽约的所罗门·R. 古根海姆博物馆(Solomon R. Guggenheim Museum)就是很好的一个例子,但是这种做法会让博物馆建筑凌驾于自身内容之上,因而显得喧宾夺主。如果一座博物馆无法拥有类似古根海姆博物馆的藏品,那么就应该设计得更低调,更具功能性。中国某些知名的博物馆就位于那些不起眼的建筑中。广州的时代美术馆位于一幢普通住宅楼的顶层,北京的尤伦斯当代艺术中心则在由旧的电子工厂大院改造而来的艺术区里面。上海玻璃博物馆并非一株含羞草。相反,它具有一个精雕细琢的外立面,日夜迎候和召唤远道而来的客人。但在建筑设计上没有高高在上的姿态,而是重建和扩展了现有结构,让建筑回归到更

好地服务于博物馆之上。

上海玻璃博物馆成功的另一个原因是它将自身定位为对特定艺术门类的展示,而非泛泛的艺术博物馆。随着中国财富的增长,对艺术品的购买也持续增加(《赫芬顿邮报》报道:"中国的超级富豪花费巨资来打造属于自己的艺术博物馆。"),但"蒙娜丽莎"和"星空"恐怕在短期内还不会来到中国。[5] 中国的博物馆如果希望在藏品方面和卢浮宫、大都会艺术博物馆、荷兰国立博物馆相匹敌,并且像它们那般宾客盈满,或许尚待时日。因此,在中国致力于建设展示特定艺术门类的新型博物馆更加现实,也更容易出类拔萃。这类具有定向内容展示的博物馆包括上海邮政博物馆、中国烟草博物馆和上海公安博物馆等,这些博物馆都有自己特定的参观人群。有些参观者甚至不去近在咫尺的艺术博物馆,却为了20世纪70年代的美国老爷车跑到远郊的嘉定参观上海汽车博物馆。同这些博物馆一样,上海玻璃博物馆也会为参观者集中呈现与玻璃有关的历史,通过策划玻璃制作与应用的展览来重现历史场景。不仅如此,通过对于玻璃艺术品的展示,也扩大了潜在观众群。

最后,上海玻璃博物馆凭借自身的专业性来实现自身定位。博物馆与住宅、办公楼或医院等那些需要在空间和功能的设置上有一定标准的建筑有所不同,博物馆并不需要这种定义极其明确的设计。许多中国的新建博物馆因为策划的薄弱而饱受诉病,另一些博物馆只是虚有其表地造了一个建筑外壳,甚至根本没有藏品。因此,这正是博物馆设计者工作的题中之

义，他们有责任通过设计来传达博物馆的策划，而他们接收的信息更关系到一个博物馆的成败。上海玻璃博物馆的建筑师与设计师对藏品、选址、老建筑、潜在活动策划进行了研究，最终完成了设计。他们所收集并研究的信息最终体现在了博物馆设计上，使得博物馆具备了天时地利的优势。

中国的新兴博物馆要取得成功，不能采取"先建了再说"的态度。相反，必须寻求一种切实可行的设计方案，而且这个方案是经过深思熟虑、综合多种因素之后所达成的结果。从这个意义上说，上海玻璃博物馆扩展了项目策划与资金来源，对于工业建筑进行重新利用的同时与周围环境相融合，在展示特定的艺术门类的同时通过研究使得决策最终落实。

事实上，中国有太多的新兴博物馆企图复制弗兰克·盖里（Frank Gehry）创作的古根海姆博物馆——一个令西班牙毕尔巴鄂这个城市名声大噪的项目，他们中有的还在尝试，而有的则已经失败。聪明的博物馆建造者意识到，在21世纪的今天要建造一所世界级的博物馆并非像单纯地建造一个地标性建筑那么简单。在中国能够取得成功的新兴博物馆，一定是那些认识到了自身在选址、展示内容、观众等方面的不足，并能利用这些限制条件使之转变为设计的可能的博物馆。

简必珂（Clare Jacobson）
《中国新兴博物馆》作者

1. Frank Langfitt 于2013年5月21日在All Thing Considered栏目上指出："中国建造了众多博物馆，但展示什么内容却又是另外一回事。" www.npr.org/blogs/ parallels/2013/05/21/185776432/china-builds-museums-but-will-the-visitors-come.

2. Jonathan Jones 于2013年7月17日在theguardian.com 上指出："中国博物馆藏品多达40 000件赝品。" www.theguardian.com/culture/2013/jul/17/ jibaozhai-museum-closed-fakes-china.

3. 杨旭于2012年5月23日的《人民日报》网上刊文称："免费参观也无法吸引人们踏入博物馆。" http://english.peopledaily.com. cn/90782/7824062.html.

4. 2013年5月15日查询韦氏第三版新国际英语大辞典（Merriam-Webster Unabridged）中对"博物馆"的注解。 http://unabridged.merriam-webster. com/unabridged/museum.

5. Kelvin Chan 于2012年5月9日在《赫芬顿邮报》刊文指出："中国的超级富豪花费巨资来打造属于自己的艺术博物馆。" www.huffingtonpost.com/2012/05/09/ china-super-rich_n_1502446.html.

INTRODUCTION

China's museum-building boom has become a popular story in the press. It is typically a sad story. In the United States, National Public Radio cries, "China Builds Museums, But Filling Them Is Another Story."[1] The United Kingdom's Guardian announces, "Scandal in China over the Museum with 40 000 Fake Artefacts."[2] China's own People's Daily states, "Free Entry Cannot Attract Visitors for Chinese Museums."[3] But there is another story of new museums in China, a story that does not make front-page news. It is a story of successful museums, quiet complements to their headline-grabbing siblings. These museums show what is going right in China's museum-building boom.

The Shanghai Museum of Glass is one of these museums. Unlike more notorious museums, it is a place that considers its content as much as its architecture, its visitors as much as its neighbors, its long life as much as its initial reception. And with these basic gestures, it is a success. This success is not unprecedented. Rather, it is based on some key components that it shares with other thriving new museums in China. In following some simple but essential guidelines, the Shanghai Museum of Glass tells a story that might not make the news, but might be the right approach in contemporary China.

First, it is a museum that is more than an exhibition space. The definition of a museum as "an institution devoted to the procurement, care, study, and display of objects of lasting interest or value" seems somewhat out of date in the twenty-first century.[4] While walls covered with paintings and display cases filled with objects are at the heart of any contemporary museum, it now needs to expand its definition in order to remain culturally and economically viable. This is true for museums around the world as it is in China, as competition from other cultural venues increases and as traditional sources of museum funding decrease. New museums in Shanghai heed this call for expanding their programming and fund raising in a variety of ways: the Rockbund Museum runs a series of evening lectures and performances, the Museum of Contemporary Art offers space for sponsored programs, and the Shanghai Nature Museum incorporates a large central garden. The Shanghai Museum of Glass has an added incentive to offer more than just exhibitions. Unlike these other Shanghai museums, it is far from the city center and well off the tourist path. Visitors who come all the way to its home in the remote Baoshan District should expect their trip to be worth the effort. To fulfill this need, the museum offers interactive exhibitions where visitors do more than look at the art. It provides a day's worth of auxiliary spaces, including a hot glass show, a café, and a DIY workshop. In addition, it offers event spaces to both extend its collection to an audience beyond the typical museumgoer and to bring added revenue to the project. Another key component of the Shanghai Museum of Glass's success is its reuse of industrial buildings. Such buildings have become popular venues for new museums across the globe. Their sturdy shells can provide the vast open spaces and abundant natural light that contemporary art requires. Museums developed in former industrial buildings are sustainable models that offer a physical connection to an area's history. At the same time, they provide a cheaper alternative to demolishing a site's structures and building from scratch. The Tate Modern in London (a former power station) and the Massachusetts Museum of Contemporary Art in North Adams (originally a cloth-printing factory) reused industrial buildings to achieve international acclaim. Closer to home, Shanghai's Power Station of Art (another former power station) and Minsheng Art Museum (a former steel factory). The Shanghai Museum of Glass, once the manufacturing headquarters of the Shanghai Glass Company, is a notable example of reuse in that the museum's art reflects the former industry of its buildings. This connection was in part mandated; land-use laws necessitated that the site would continue to have glass-related use after most of the glassworks moved offsite. Still, it is somehow exceptional that the old manufacturing buildings have a direct relationship to the new museum, a relationship that is employed in the museum design and programming.

Hand-in-hand, with this, the Shanghai Museum of Glass is successful in that it does not attempt to be an icon. While the building certainly is designed to be a centerpiece of its larger site and of the Baoshan District in general, it is meant more to fit into its context than to stand out as a singular formal statement. Iconic buildings do not necessarily make bad museums; consider New York's Solomon R. Guggenheim Museum. But they do necessarily prioritize the building over its contents. And if a museum is not privileged to own a Guggenheim-style collection, its building needs to take a bit of a back seat in order to make a fully functioning project. Some of China's best-regarded museums exist in rather submissive buildings. The Times Museum in Guangzhou sits at the top of a nondescript residential tower, while the Ullens Center for Contemporary Art in Beijing is one of many art centers in a cluster of former electronics factories. The Shanghai Museum of Glass is no shrinking violet. Its intricate facade beckons, day or night. But in minimizing its "look at me" architectural design with its reuse and expansion of extant structures, it allows the buildings to truly work as a museum.

Another key component of the success of the Shanghai Museum of Glass is that it showcases a specific kind of art, rather than positions itself as a general art museum. While Chinese money is purchasing fine art at an ever-increasing rate (the Huffington Post claims "China Super Rich Use Boom Money to Open their Own Art Museums"), the Mona Lisa and The Starry Night are not coming to China anytime soon.[5] It will take some time before the collections of Chinese museums can compete with those of the Louvre, the Metropolitan Museum

1. Frank Langfitt, "China Builds Museums, But Filling Them Is Another Story," All Thing Considered, May 21, 2013. www.npr.org/blogs/parallels/2013/05/21/185776432/china-builds-museums-but-will-the-visitors-come.

2. Jonathan Jones, "Scandal in China over the Museum with 40 000 Fake Artefacts." theguardian.com, July 17, 2013, Jonathan Jones. www.theguardian.com/culture/2013/jul/17/jibaozhai-museum-closed-fakes-china.

3. Yang Xu, "Free Entry Cannot Attract Visitors for Chinese Museums." People's Daily Online, May 23, 2012. http://english.peopledaily.com.cn/90782/7824062.html.

4. "Museum" Merriam-Webster Unabridged, accessed October 15, 2013. http://unabridged.merriam-webster.com/unabridged/museum.

5. Kelvin Chan, "China Super Rich Use Boom Money to Open their Own Art Museums." Huffington Post, May 9, 2012. www.huffingtonpost.com/2012/05/09/china-super-rich_n_1502446.html.

of Art, the Rijksmuseum, and the like to bring in visitors. Focusing a new museum in China on a specific kind of art can be a more realistic and viable way to make it stand out from the crowd. Content-specific museums in Shanghai range from the Shanghai Postal Museum to the China Tobacco Museum to the Shanghai Museum of Public Security. Each has its own audience. People who might never consider going to the art museum around the corner might travel out to the distant Anting District to see the Shanghai Auto Museum's collection of 1970s American muscle cars. Like these museums, the Shanghai Museum of Glass shows visitors a history of the subject in focus. It incorporates exhibits on making and using glass to enliven that history. In addition, it displays fine art made from glass, thus broadening its potential audience.

Finally, the Shanghai Museum of Glass uses design expertise to inform its choices. Unlike a commission for a house or office tower or hospital, which requires a standard set of rooms and features, that for a museum does not necessarily have a well-defined program. Many new museums in China have been criticized for their lack of program, many others for being only building shells without a collection at all. It is the responsibility of museum designers to help inform the project's program, and museums can succeed or fail depending on the information they receive. The architects and exhibition designers of the Shanghai Museum of Glass researched the collection, the museum site, the extant buildings, and potential activities to drive their design. The information they gathered led to the reserved approach that the museum takes, an approach that makes good sense in its place and time.

For a new museum in China to succeed, it cannot follow the "build it and they will come" model. Instead it needs to recognize that a viable design is one that incorporates many well-considered components. At the Shanghai Museum of Glass, broadening the programming and fund raising, reusing industrial buildings, designing within the neighborhood context, showcasing a specific subset of art, and using research to inform decisions all assist in its realization.

It is true that too many new museums in China are trying, and failing, to duplicate the notoriety that Frank Gehry's Guggenheim Museum, Bilbao so famously brought to the Spanish city. Smarter museum makers recognize that creating a world-class museum in the twenty-first century is not as easy as building an iconic piece of architecture. The successful new museum in China will be one that recognizes the limitations of its site, content, and audience and uses those limitations as design possibilities.

Clare Jacobson
author of *New Museums in China*

城市规划 URBAN PLANNING

地理位置

上海玻璃博物馆坐落于长江西路685号, 这里是上海轻工玻璃公司宝山厂区的旧址。宝山区位于上海市区外围, 距市中心约15公里, 曾是上海著名的重工业基地。值得一提的是, 全球第二大钢铁生产商宝钢集团正是因宝山区而得名, 宝钢的一个厂房就位于上海玻璃博物馆正后方。

宝山区的工业环境不甚理想, 休闲文化领域的发展也相对滞后。博物馆周边充斥着新旧厂房、货运集装箱、工业园区及仓库, 满是旧时工业生产的痕迹。尽管近几年已开始向第三产业转型过渡, 但这一区域仍给人工业基地那种偏安一隅的印象。一些旧厂房如今成为了当地工人的临时居所或临时商铺。

但曾经步履蹒跚的宝山区正经历着转型与巨变, 逐渐进入后工业时代。上海大学在20世纪90年代中期将主校区迁至此处, 几个现代化办公园区也悄然兴起。园区计划以新兴产业为依托, 竭力吸引大型研发项目入驻, 一座座办公与住宅综合体拔地而起。通往市区的地铁1号线和3号线已通车多年, 另有两条地铁线规划于2020年建成。基础设施、住宅和商业项目的开发, 带动了区域品质的整体提升, 宝山区将成为上海又一个乐享城市转型改造丰硕成果的行政区块。

宝山区日新月异的喜人变化, 正是玻璃博物馆选择上海轻工玻璃公司旧址的用心所在。21世纪初期, 上海轻工玻璃公司终止了该厂原规模化生产功能, 将其改造转变为工业园区, 面向其他低端制造及仓储业态提供租赁服务, 因而留下了这块空地, 如今正好为博物馆提供了展现魅力的充足空间。宝山区政府亦对将这块场地用于文化项目开发倍加赞赏, 认为这对于整个区域的重新定位意义非凡。由此可见, 看似偏远的长江西路685号绝对称得上是上海玻璃博物馆的最佳馆址。

LOCATION

The Shanghai Museum of Glass is situated at 685 West Changjiang Road, on the site of a former glass factory owned by the Shanghai Glass Company Ltd. in Baoshan. Located approximately 15 kilometers north of Shanghai's city centre, Baoshan is one of Shanghai's outer districts and formerly a hub for heavy industry. In particular, the area lent its name to state-owned company Baosteel, the world's second largest steel producer. One of the group's manufacturing halls is located directly behind the Shanghai Museum of Glass.

The urban landscape surrounding the museum is predominantly industrial in character, comprising old and new factories, shipping containers, industrial parks and warehouses. The area feels remote, with a reputation for manufacturing still standing strong despite a shift in recent years to tertiary industries. With few redeeming qualities, Baoshan's industrial fringe offers little in the way of leisure or cultural attractions. Instead, it is known for a string of disused factories, many of which function as informal temporary accommodation for locally-based workers, or are reused for short-term commercial endeavours.

Yet Baoshan is undergoing a transformation, and is moving into a post-industrial era. Shanghai University relocated their main campus here in the mid-1990s, and construction has also begun on several modern office parks. Aimed at targeting new industries, some house research and development facilities. Residential complexes are also being built across the district. Already connected to downtown Shanghai via metro lines one and three, and with two further lines scheduled for completion by 2020, the suburb is on the cusp of change. With improved infrastructure, and residential and commercial developments all poised to raise the general quality of the area, Baoshan is one of Shanghai's next suburbs set to benefit from urban regeneration.

The transformation of the area was a leading factor in the decision to locate the Shanghai Museum of Glass in the Shanghai Glass Company Ltd.'s Baoshan facility. A shift in focus from large-scale production to low-level manufacturing and storage in the early twenty-first century had left behind a vacant site, offering a spacious location for a museum. Cultural redevelopment of the site was considered a welcome contribution to the wider repositioning of Baoshan district by local authorities. Despite its remote location, 685 West Changjiang Road was favoured as the location for the Shanghai Museum of Glass.

现场情况

上海玻璃博物馆占地面积约为40 414平方米，原建筑面积27 920平方米，共有31栋建筑物，包括许多小型临时结构建筑。多年以来，人们在老建筑之上又陆续进行了多次加盖，既有木屋顶的砖墙结构，也有较为现代的混凝土结构。其中建筑物的修建横跨35年。建筑群中最早的建筑要追溯到1958年，改造将热玻璃表演厅巧妙地设于其中，而最新的建筑则到1993年才完工。

SITE DESCRIPTION

The site on which the Shanghai Museum of Glass is situated covers a total of 40 414 sqm, with an original gross floor area of 27 920 sqm. The site counted thirty-one buildings, including many small-sized temporary structures. Over the years, multiple structures were added to the site on demand, varying from brick structures with wooden roofs, to more contemporary ones in concrete. Erected over a period of thirty-five years, the oldest, currently housing the Hot Glass Performance Hall, dates from 1958, while the most recent was finalized in 1993.

在场地中央2 925平方米的区域上，引人注目的博物馆大楼是建筑群中最为核心的结构。该楼建造于1991年，高20.2米，檐高15.5米，采用混凝土及钢结构窗框，主馆大楼由敞开式大厅与二楼画廊组成。两个副楼分别位于北侧与西侧，此前用于办公和仓储。

尽管别具匠心的选址令建筑具备了厚重的历史感，并与博物馆主题高度契合，但是旧址建筑本身的利用价值极其有限。由于一直缺乏良好规划，导致原先的建筑密度较高，视野不够开阔。博物馆主楼前的小型楼宇和底层商铺将主楼与长江西路阻断，不仅遮挡了视线，还令建筑与市政主干道无法直接连通。整个区域内仅开辟有一条位于建筑群侧面的狭窄小路，与博物馆建筑还有一定距离。

At 2 925 sqm and positioned at the centre of the site, the designated museum building formed one of the largest structures of the cluster. Built in 1991, it extends to 20.2 metres in height with eaves positioned at 15.5 metres. A concrete structure with steel frame windows, the main body of the building comprises an open-volume hall with a second floor gallery. The construction features two annexes to the north and west, originally used for office and storage space.

Although its historical ties to the Shanghai Glass Company Ltd. thematically made for an interesting location for the Shanghai Museum of Glass, the original site offered little in architectural value. A lack of planning over the years had resulted in a high building density and few clear sight lines. In addition, a smaller structure and precinct of low-level shops in front of the building designated as the main museum blocked the site from the main road. The physical obstructions complicated both visibility and access from West Changjiang Road, with the entire site served by a narrow service road positioned to one side of the grouping of buildings, some distance from the designated museum structure.

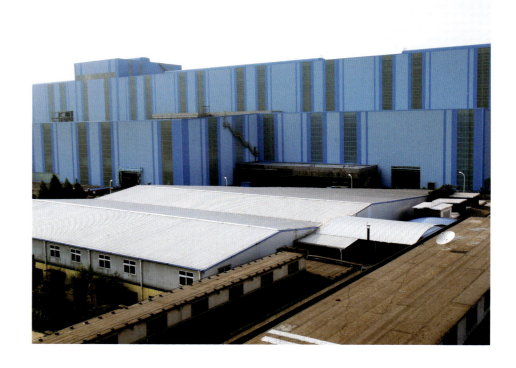

项目定位

在地理位置和现场条件的双重挑战之下，玻璃博物馆亟需行之有效的定位策略。尽管在选址方面拥有着深厚历史渊源，但宝山区并非传统意义上的文化区域。作为远离市中心的典型工业基地，这里在休闲文化方面乏善可陈，很难集聚人气。常驻居民较少、公共交通不够便利 (从市中心到博物馆所在地最快也需45分钟车程) 等诸多不利因素在很大程度上影响着上海玻璃博物馆的参观人流。

PROJECT POSITIONING

The dual challenge of both location and site formed the core of the project's positioning strategy. Despite the site's historical significance through its former role as a manufacturing facility for Shanghai Glass Company Ltd., its location in Baoshan was off the beaten track for cultural purposes. Set deep within the industrial fringe of one of the city's most remote suburbs, an absence of noteworthy leisure or cultural opportunities mean few tourists venture to the area.

Similarly, few downtown residents frequent Baoshan, with other areas of the city offering more attractive alternatives for relaxation, education and spending time with family and friends. With public transport routes still in development, and the most convenient method of reaching the museum from the city centre being a 45-minute car journey, the location experiences a low-level of foot traffic, making spontaneous visits to the Shanghai Museum of Glass unlikely.

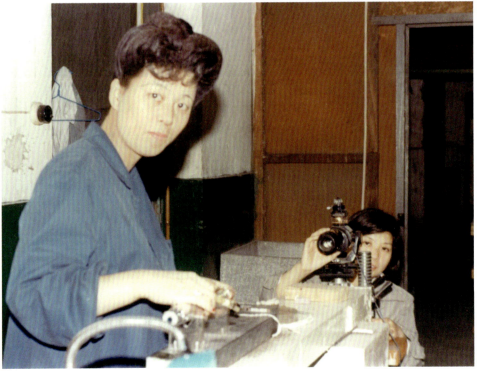

1　游客中心
　　Visitor Centre

2　上海玻璃博物馆主场馆
　　Museum Main Hall

3　珂庐私人会所,
　　设计品商店, 伽蓝画廊
　　Private Members' Club KILN,
　　Design Store, Glaze Gallery

4　热玻璃演示厅
　　Hot Glass Performance Hall

5　博物馆咖啡店
　　Coffee Shop

6　DIY 创意工坊
　　DIY Creative Workshop

Masterplan
1/1500
总图

N

作为私营机构的上海玻璃博物馆主要依靠门票销售及各类活动的营收。对于一家成功的博物馆来说，人气决定一切，只有源源不断的人流才能打造出活力四射的文化品牌。由于有国际成功案例的珠玉在前，玻璃博物馆亦着力为多次莅临参观的宾客量身打造优质服务空间，让博物馆成为他们不断学习、积极互动和获取灵感的不二之选。因此，项目定位不仅要将吸引力放在首位，还需制定可持续发展策略，用心培养地理品牌概念。

提升人气绝非易事。简单的玻璃艺术展示显然无法说服人们不顾舟车劳顿欣然前来参观，层出不穷的特色呈现与独具创意的附加功能才是吸睛的必胜法宝。只有具备了令人折服的独特魅力，玻璃博物馆才能在上海乃至全国的众多文化去处中脱颖而出。在宝山区高举转型大旗的背景下，加大投资将博物馆打造成为积聚人气的文化品牌，无疑能够极大地推动这一城郊结合部从工业集散地向时尚生活中心转变。上海玻璃博物馆必须成为后续项目的一面旗帜，传递出宝山区全新的文化面貌与形象特质。

A privately-funded venture, the Shanghai Museum of Glass' success is linked closely to ticket sales and revenue from activities and events. Related, a shared goal of both the designers and client was to create an attractive, 'living' cultural hotspot for Baoshan. Taking inspiration from international case studies, the aim was set to build a community of visitors that ideally pay not single, but multiple visits, and consider the museum a place to continuously learn, interact and be inspired. The project positioning therefore not only centres on attracting visitors; it also offers sustainable strategies to generate awareness of the intention of the location.

The challenge of driving visitors to the venue points to the need for a destination museum that surpasses expectations. Firstly, it needed to be of sufficient size to sustain visitors' attention for a period of several hours to justify travel time. To that end, the housing of unique attractions and additional functions to elevate the venue beyond a straightforward showcase for glass art was pivotal in terms of competing with comparable cultural destinations in Shanghai and China. Longer term, the venue is positioned as a 'loss leader' for the wider redevelopment of Baoshan, spearheading the suburb's shift from industrial hub to lifestyle centre. To set a precedent for this, the venue needed to brand the area for subsequent projects with a strong visual identity to indicate change.

第二类博物馆

应接不暇的各项挑战促使上海玻璃博物馆走向了"第二类博物馆"的设计路线。大量研究显示，符合国际标准的玻璃博物馆大致可以分为第一类或第二类。第一类博物馆一般规模较小，倾向于围绕单一的藏品设计。这类博物馆通常能够成为拥有特定兴趣及知识储备的参观者们趋之若鹜的朝圣之地。它们往往只有一到两个展厅，参观者的平均逗留时间约为30分钟。

而第二类博物馆则规模较大，强调功能的多元化，打破传统博物馆的单一理念，追求兼收并蓄的文化内容及品位休闲的运营模式。美国的许多玻璃博物馆便是这一类型的杰出代表，如纽约康宁玻璃博物馆 (Corning Museum of Glass)、托莱多艺术博物馆内的玻璃展馆 (Glass Pavilion at Toledo Museum of Art) 以及塔科马的玻璃博物馆 (Museum of Glass in Tacoma)。第二类博物馆非常注重参观者的体验，而不是简单地积累和展示藏品。展区设计必须考虑多种目标人群的参观需求。博物馆亦会提供各类配套设施——DIY创意工坊、商店、咖啡馆和活动空间等，这样既可在门票销售的基础上开拓其他营收手段，也能让参观者的"发现之旅"更为精彩丰富。

THE TYPE TWO MUSEUM

The various challenges posed by the venue's location called for the Shanghai Museum of Glass to develop itself as a 'Type Two' museum. Extensive research revealed that internationally, glass museums can be categorized as either a Type One or Type Two destination. Type One museums are generally small in scale, and tend to centre around a single collection. As such, they are effective in attracting a special-interest audience. Often occupying just one or two rooms, their small size results in average visitor stays of approximately 30 minutes.

In contrast, Type Two museums are larger in size, and combine a collection with additional functions and activities, stretching the concept of the traditional museum towards an all-encompassing leisure destination. A popular approach for US-based museums of glass in particular, they include New York's Corning Museum of Glass, the Glass Pavilion at Toledo Museum of Art, and the Museum of Glass in Tacoma. This type of museum places a greater emphasis on the visitor experience instead of the collection. Exhibits are designed to engage a multitude of target groups. In addition, they offer a range of facilities such as workshops, stores, cafés and event spaces to provide an alternative source of revenue separate from ticket sales, and resulting in visitor stays of several hours at a time.

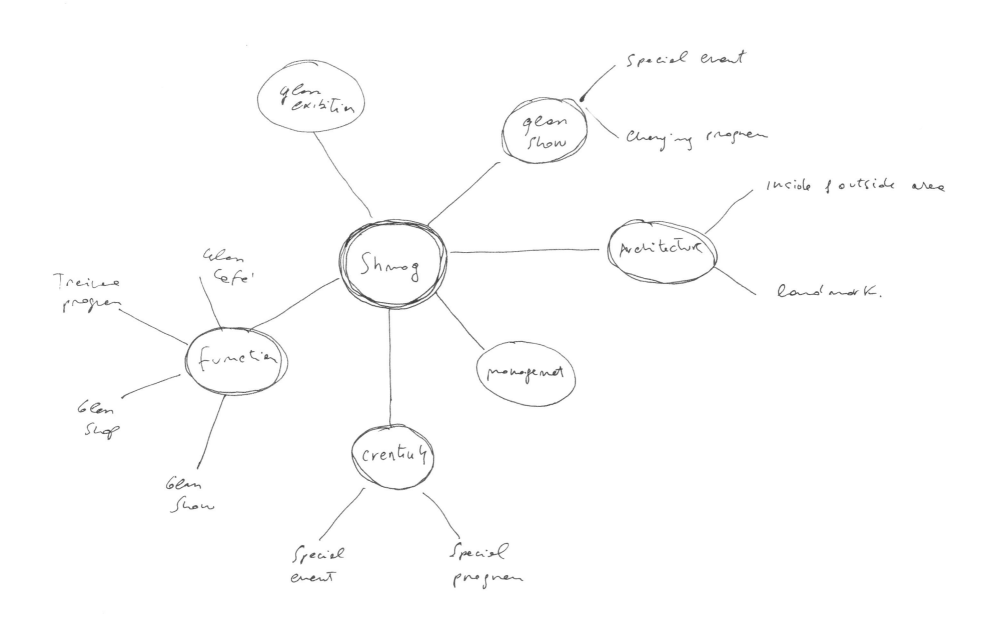

亲子活动是这类博物馆的一大亮点，通常采用寓教于乐的方式，非常受欢迎。与侧重文字的静态展示方式相比，互动展示模式更能鼓励孩子们亲身动手操作，调动他们参与的积极性。青少年无疑将带来无限活力，令博物馆博采众长、生动有趣的特质得以彰显，从而进一步担负起人才培养的社会责任。

为了在众多休闲活动中脱颖而出，进一步打造自身竞争优势，第二类博物馆设计了夺人眼球的多项体验活动，将惊喜不断的各类演出要素贯穿其中，为参观者营造出充满活力的学习与娱乐空间。特色体验活动包括了DIY创意工坊、炫动表演、经典展示及专家授课。国际上的众多玻璃博物馆都设计有热玻璃表演环节，观众可在此亲身体验玻璃吹制过程。表演融合火焰、蒸汽及高温等日常生活中罕有的元素，自始至终都能让人屏息凝神、全神贯注，从而给予观众，尤其是儿童参观者留下深刻印象。

第二类博物馆在中国尚属全新的品牌理念，亦是上海玻璃博物馆借以吸引人气的全新尝试。将博物馆定位为融合创意工坊及特色表演的休闲场所，无疑更能引起人们的广泛兴趣。体验与动感元素的巧妙结合颇受年轻家庭的青睐。中国市场上年轻家庭的休闲娱乐消费能力正在快速增长，而第二类博物馆正好迎合了这一群体合家文化活动的需求。

Generally adopting an approach of learning through doing, Type Two museums are most popular amongst families with children. They successfully engage with younger audiences via an accessible approach, including hands-on and interactive exhibits as opposed to static, text-heavy displays. By extension, young visitors bring to these venues an attractive energy, further reinforcing concepts of an active, living museum dedicated to the nurturing of new talents.

To compete with their respective leisure-based alternatives and to further encourage visitor stays of up to several hours at a time, Type Two museums incorporate 'live' elements and experiences that effectively shift the museums' approach from passive to active, presenting themselves as living spaces for learning and entertainment. These include DIY workshops, shows, demonstrations and lectures. A shared characteristic of all the international case studies is a hot glass show, where audiences experience first hand the skill, dexterity and spectacle of glass-blowing. Incorporating the thrill of fire, steam and intense heat, the displays capture visitors' attention and create a lasting impression, particularly amongst children.

A brand new concept in China, the Type Two model offered an ideal way for the Shanghai Museum of Glass to engage visitors for stays of up to several hours. Positioning the museum as a leisure destination that encompasses additional facilities, workshops and demonstrations not only attracts visitors from downtown areas and beyond, but also justifies travel time. The strong emphasis on experience, partnered with 'live' elements, appeals specifically to young families. In China, this target group now possesses increasing budgets for leisure, but experiences a lack of choice in day-filling activities that combine education and entertainment. The Type Two museum therefore poses an alternative activity for families to spend time together.

总体规划的调整：
博物馆的扩展

上海玻璃博物馆规划为第二类博物馆，融合了诸如咖啡馆、礼品店、活动区等场所及特色表演等要素，力求打造令人愉悦的活力文化空间。原本2 925平方米的博物馆大楼空间过于狭小，无法容纳规划中的诸多设计理念。为了增大主馆规模，博物馆决定将最初设计的大楼与南侧632平方米的旧车间打通，并在连接走廊上方安装了玻璃屋面。玻璃贯穿两侧形成了闭合结构，一个玻璃中庭得以呈现出来。这样的设计不仅巧妙地将两个建筑物合二为一，还在两者之间开辟了崭新的室内空间。

ADJUSTING
THE MASTERPLAN:
EXTENDING THE
MUSEUM

The Shanghai Museum of Glass set out to become a Type Two museum, incorporating various facilities such as a café, shop and events area, as well as unique performance elements to bring the venue to life. The 2 925 sqm building that was designated for the museum proved too small to create an attraction of sufficient depth and appeal, and lacked adequate space to house necessary additional elements.

To increase the size of the main museum hall, the original designated building was merged with a 632 sqm former workshop located directly to the south. To unite the two, the alley that ran between the buildings was covered with a glass roof and boxed in on either side with glass to create an atrium. The solution not only successfully linked the buildings, it also resulted in an additional, new indoor space between the two old structures.

　　最初的规划方案计划将博物馆主楼东面的3栋建筑物用作活动空间，其中1958年建成的二层办公楼面积为865平方米，如今部分设计成了DIY创意工坊；1971年完工的一层厂房成为了博物馆的一部分；场地内历史最悠久、始建于1958年的厂房，目前用作博物馆的热玻璃演示厅。原有的木屋顶、砖体结构三层建筑面积达3 216平方米，其前侧的副楼在改造中几乎都被拆除，从而在主楼前形成了一个小广场。为了适应新的使用要求，设计对该建筑结构进行了改造，移除了二楼和三楼，形成了开放空间。

To the east of the main museum hall, three further structures were integrated into the project's initial master plan to create venues for the additional activities that would create a Type Two destination. A two-storey, 865 sqm former office building dating from 1958 was partly repurposed as the DIY Creative Workshop. Additional original structures partially integrated into the program include a single-level factory built in 1971, and the site's oldest construction, a 1958 factory building, now used to house the museum's Hot Glass Performance Hall. A brick structure with a wooden roof, it originally spanned three levels and 3 216 sqm. The building had several annexes, the frontmost of which was demolished in order to create a small plaza in front of the building. To render the structure fit for its new purpose, the second and third floors were removed to create an open-volume space.

不同单元彼此以及与主馆之间的联系都通过博物馆的轴线设计来实现。主轴线是一条贯穿南北的人行道，两侧分布着主馆、DIY创意工坊、热玻璃演示厅、礼品店、珂庐私人会所以及Keep it Glassy国际创意玻璃设计展区。轴线将博物馆各区域串联，给人以整齐、统一的简约视感。从使用层面看，它也将原本凌乱的空间及狭窄的通道加以重新规划，形成了有机的整体。

从美学层面看，博物馆轴线的引入令改建后的工业厂房之间拥有了广阔的活动空间和绚丽的灯光效果，令场馆两端的视觉效果一气呵成。空间的多样性设计中包含了露天区域规划，使整个项目方案更显完备，亦加强了博物馆作为生活时尚中心的实用性和功能性。

Connecting the various added units both to each other and to the main museum hall is the Museum Axis. Extending from north to south, this paved walkway is flanked by the main museum hall, DIY Creative Workshop, Hot Glass Performance Hall, design store and private members' club KILN, as well as exhibition space Keep it Glassy. The axis creates a sense of unity across the museum's various locations. On a more fundamental level, it brings structure to a complex site, previously characterized by unorganized spaces, narrow alleys and blind corners.

On an aesthetic level, the Museum Axis injects spaciousness and light between the renovated industrial buildings, allowing visitors a clear line of vision from one end of the museum park to another. Offering possibilities for an open-air programme through spatial diversity, it brings cohesiveness to the development, linking the functions that elevate the Baoshan destination to a versatile leisure park.

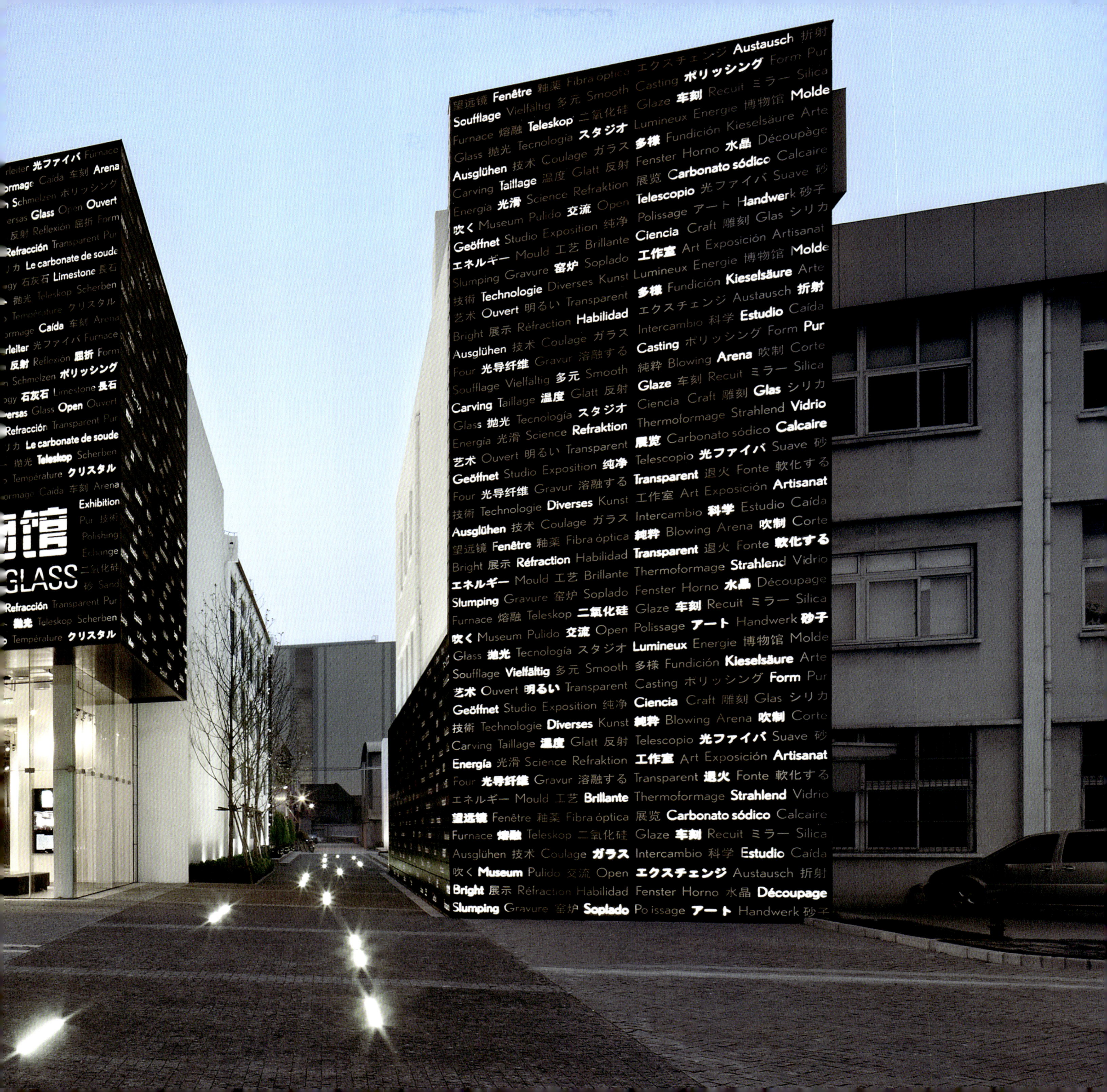
Exhibition
GLASS

总体规划的调整与整合成功地扩展了博物馆的空间，遵循了第二类博物馆的定位策略。不过，博物馆建筑的整体外部可见度还不够高。要令博物馆第一时间从长江西路上跃入人们的视线，必须对场馆入口进行调整，从而为博物馆成为宝山区域名片创造条件。

横亘在长江西路和博物馆之间的是两栋二层建筑，分别为一个车库和一个车间，属于第三方所有。最彻底的解决办法是买下这两栋建筑，将其拆除，进而打开一个40米进深的广场，这也是确保博物馆改造成功的最理想条件。露天广场形成之后，不仅使博物馆外部环境豁然开朗，同时也能提高建筑的整体质感，显得更为高雅大气。

最终将博物馆打造成令人瞩目的地标性建筑还需通过扩建1 198平方米面积的新增建筑结构，拓展博物馆主楼的范围。入口延展12.5米与广场相通，令博物馆更接近街道，且有利于塑造别具一格的外立面，使其在全球范围树立自己独一无二的品牌与形象。

The adjustment of the master plan and integration of an additional structure successfully extended the main museum hall, in keeping with its Type Two strategy. However, the site still lacked pivotal visibility from the street. A drastic intervention was required in order to render the museum easily seen from West Changjiang Road, and equally important, to create an instantly identifiable icon that would go on to act as a visual identifier for the area.

Two low-level buildings blocked the site's access and visibility from West Changjiang Road. The structures obscuring the museum, a garage and workshop, belonged to third parties and needed to be demolished in order to make the museum visible from the street, and to create a 40-metre-deep plaza. A radical measure, purchasing the buildings was deemed the only possible course of action to ensure the museum's success. The open plaza that was created not only allows the architecture to be seen by passers-by, it also 'frames' the museum buildings, imbuing them with a sense of dignity and grandeur.

The final step in creating a striking and instantly recognizable landmark destination was to further extend the physical area of the main museum hall through the construction of an entirely separate new build structure, covering an area of 1 198 sqm. As well as shifting the museum's entrance 12.5 metres further into its plaza, bringing it closer to the street, it also created the possibility of creating the iconic façade that would go on to visually brand the project on a global scale.

博物馆建筑 MUSEUM ARCHITECTURE

适应性再利用

上海玻璃博物馆是一个适应性再利用项目：将玻璃厂旧址改建为多功能地标性博物馆。适应性再利用是指重新利用现有老建筑，并使其适应新功能。

在上海，适应性再利用是创意人士对办公室或工作室等场所采用的常见处理方式。如卢湾区的8号桥综合园以及红坊雕塑公园的创意中心都是创意产业界的成功典范。这类项目大多位于废弃的工厂内，与新建的现代办公楼相比，特点鲜明、个性独具。创意中心的形成可以同时起到支持创意产业和保护上海工业遗产的双重作用。在上海以外，原为军工厂的北京798艺术区、德国艾森的关税同盟煤矿工业区 (Zollverein Coal Mine Industrial Complex) 以及最终定址于美国纽约麦迪逊大道，原爱迪生联合电力公司 (Consolidated Edison) 配电站的安迪·沃霍尔 (Andy Warhol) 标志性"工厂" (Factory)，都可谓适应性再利用的经典之作。艺术、文化、创意与工业空间再利用之间的密切联系在这些项目中体现得淋漓尽致。

ADAPTIVE
REUSE

The Shanghai Museum of Glass is an adaptive reuse project: former glass manufacturing buildings have been converted into a multifunctional landmark museum. Adaptive reuse is defined by the reutilization of existing buildings, and their adaptation for functions other than their original intended use.

In Shanghai, adaptive reuse is an approach most commonly used for creative hubs: clusters of offices or studios, popular amongst the creative industries, and including Luwan district's Bridge 8 complex, and Red Town's sculpture park. Generally housed in abandoned factories, they offer a unique character not found in modern, purpose-built office buildings. In this way, they serve a dual function of supporting creative businesses, and helping to preserve Shanghai's industrial heritage. Beyond Shanghai, examples include the galleries of Beijing's 798 Art Zone housed inside former military factories; Zollverein Coal Mine Industrial Complex in Essen, Germany; and indeed the final location of Andy Warhol's iconic 'Factory' inside an old Consolidated Edison substation on Madison Avenue in New York. All point to a relationship between art, culture, creativity and reused industrial spaces.

工业建筑的适应性再利用理念在国际上被广泛认可, 在20世纪90年代末进入中国, 起初发展非常缓慢, 最近几年开发商们才逐渐意识到其巨大的商业潜质、文化价值及投资回报。上海玻璃博物馆在践行这一理念的同时, 将建筑的功能性拔到更高层次, 即对原有厂房重新定位, 使其成为城市中最受瞩目的建筑类型——博物馆。

适应性再利用创造了诸多优势。首先从操作层面看, 它一方面比一座全新设计、规划和建造的建筑物更为快捷, 另一方面也缩短了审批时间; 其次, 适应性再利用能够延长原有建筑的使用寿命、减少垃圾掩埋, 是一种更为绿色、更具可持续潜力的选择。尤其对于像宝山项目现场的这种混凝土结构建筑来说, 扩展建筑结构的适用性之后, CO_2密集型建筑材料便可得到更为有效的使用。

A widespread practice internationally, adaptive reuse of industrial buildings was introduced to China in the late nineties. The practice has had a slow start in China, with developers only recently realizing the commercial and cultural value adaptive reuse presents, and the fast return on investments such projects offer. The Shanghai Museum of Glass continues this concept, taking the practice to a new level through the repurposing of a former manufacturing plant for one of the most highly-regarded building types in cities: a museum.

The approach brings with it several advantages. On a practical level, it poses a considerably faster alternative to the designing, planning and erecting of an entirely new structure, as well as a shorter approval process in China. Secondly, by extending the lifespan of the original buildings and reducing landfill use, adaptive reuse offers a greener, more sustainable option. In the case of concrete buildings in particular, like those at the Baoshan site, extending the usability of structures makes more efficient use of their very CO_2-intensive building materials.

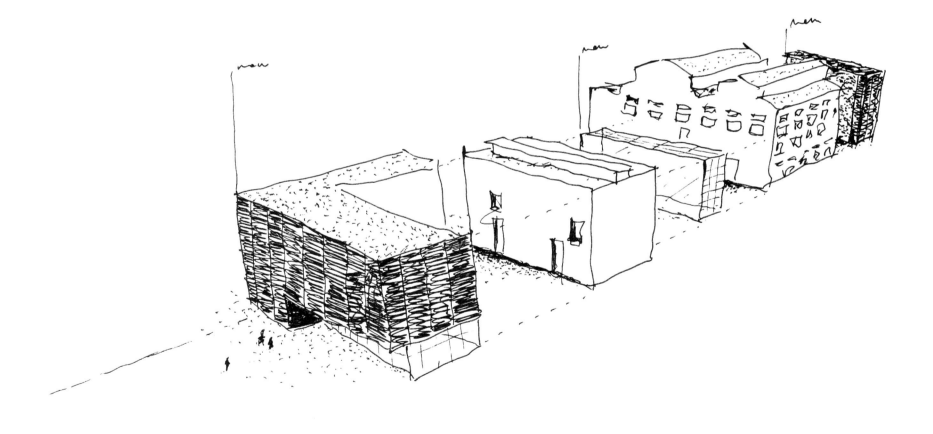

但总体来说，上海轻工玻璃厂旧厂房的原有建筑并不能充分展现适应性再利用的明显优势。为了塑造出一个能挖掘并反映原有建筑特点的博物馆，现有建筑在重新定位时，保留了以往上海工业建筑的特点，同时也快速驱动了宝山这个较为偏远地区的城市转型进程。

要使现有结构符合博物馆，尤其是作为玻璃博物馆的要求，必须进行一系列调整。设计充分利用内部空间进行展示，将这些曾经代表着地块特点的平庸建筑集中整合，通过融入当代建筑要素赋予其全新面貌，同时达到满足其使用目的的要求。最终设计不仅解决了空间和功能的问题，还大胆使用极具特色的玻璃外立面，为这一区域打造出独一无二的地标性建筑。不仅如此，重新定位无疑成为贯穿上海玻璃行业过去、现在与未来的纽带，这又与博物馆的核心理念不谋而合。

Despite the general advantages of the adaptive reuse model, the Baoshan site offered little architectural value, with the significance of the architecture linked solely to the buildings' original function as the former manufacturing premises of the Shanghai Glass Company Ltd. In creating a museum that celebrates and explores the very same medium the site once produced, the repurposing of existing buildings here serves to preserve elements of Shanghai's industrial past and kick-start urban regeneration in Baoshan's outlying areas.

Significant adjustments were necessary in order to render the existing structures suitable for a museum dedicated to glass. Although they were made efficient use of in the housing of exhibits, the unremarkable buildings that once characterized the site take second place to altogether more eye-catching contemporary interventions that bring the site up-to-date and fit for purpose. The resulting design not only solves problems of space and functionality, it also allows freedom to create a unique landmark for the area by way of a striking glass façade. More than that, its repurposing presents a direct link between the past, present and future of glass in Shanghai, mirroring a core concept of the museum.

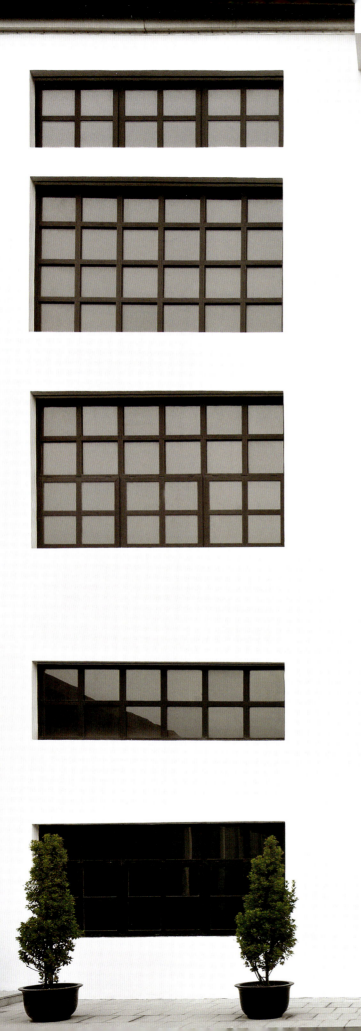

Presionando Cutting Tallado Corte プレッシング Blasen
切削 Pressing Carving 切割 Schneiden Pressage 圧制
Soufflage Drücken Blowing 吹制 Soplado Casting 吹く
Taillage 铸造 Offen Fusion Einformen Coulage Fundición
Schmelzen 熔融 Polissage ポリッシング Polishing 能量
抛光 Pulido Polieren Energy Studio Temperature Estudio
Presionando 温度 Échange Energie エネルギー Energía
工作室 Kunst Température スタジオ Art Habilidad 车刻
クラフト Technologie 艺术 Handwerk Artisanat アート
工艺 Technology オープン Open 技術 Tecnología 技术
切削 Pressing Carving 切割 Corte プレッシング Blasen
Einformen Coulage Fundición Presionando Cutting Tallado
ポリッシング Polishing 能量 Soufflage Drücken Blowing
Grabado 车刻 Gravur 软化 Taillage 铸造 Offen Fusion
Schneiden Pressage 圧制 クラフト Technologie 艺术
Energie エネルギー Energía 软化する Temperatura Four
Studio Temperature Estudio Erweichen Caída Slumping
Soufflage Drücken Blowing Schmelzen 熔融 Polissage
スタジオ Art Habilidad 车刻 工作室 Kunst Température
Corte プレッシング Blasen 切削 Pressing Carving 切割
Engraving Découpage 彫刻 Presionando 温度 Échange
抛光 Pulido Polieren Energy 工艺 Technology オープン
ポリッシング Polishing 能量 Presionando Cutting Tallado
Handwerk Artisanat アート Studio Temperature Estudio
Soufflage Drücken Blowing Soufflage Drücken Blowing
クラフト Technologie 艺术 スタジオ Art Habilidad 车刻
Erweichen Caída Slumping Grabado 车刻 Gravur 软化
Open 技術 Tecnología 技术 Engraving Découpage 彫刻
Einformen Coulage Fundición 抛光 Pulido Polieren Energy
Taillage 铸造 Offen Fusion Schneiden Pressage 圧制
工作室 Kunst Température ポリッシング Polishing 能量
Schmelzen 熔融 Polissage 软化する Temperatura Four
Grabado 车刻 Gravur 软化 Energie エネルギー Energía
Handwerk Artisanat アート 抛光 Pulido Polieren Energy
Presionando 温度 Échange Soufflage Drücken Blowing
クラフト Technologie 艺术 工作室 Kunst Température
Presionando Cutting Tallado 切削 Pressing Carving 切割
Soufflage Drücken Blowing Presionando 温度 Échange
Energie エネルギー Energía Erweichen Caída Slumping
スタジオ Art Habilidad 车刻 Schmelzen 熔融 Polissage
Corte プレッシング Blasen Einformen Coulage Fundición
工艺 Technology オープン Engraving Découpage 彫刻
Studio Temperature Estudio Taillage 铸造 Offen Fusion
软化する Temperatura Four Schneiden Pressage 圧制

博物馆大楼

上海玻璃博物馆的核心区域从长江西路入口开始由南向北延伸。建筑主要由5个部分组成：新建的L形大楼、老车间建筑、中庭、原玻璃厂房以及重新设计的副楼。

博物馆大楼通过窗户等元素保留了原有建筑的特点，展现了大空间设计开放、通透的效果。方案于原工业建筑中融入现代要素——外立面、中庭和副楼，而空间在依旧保留工业基调的同时强化了时代感，以最终呈现出中国全新类型的博物馆形象。活力动感、历久弥新、层次分明的空间效果，令过去与现在的不同功能都得以在博物馆中体现，同时也为城市提供了一个优质的适应性再利用案例。

设计完全贴合适应性再利用模式，将新建筑整合到塑造第二类博物馆所需的功能中，并为众多延伸活动创造空间。为了充分体现新旧用途间一脉相承的关系，整座博物馆外采用极具现代感的玻璃外立面，确保博物馆从长江西路一端直到最南面都具备高度辨识性。正前方的醒目区域，同时也是L形建筑的一部分，并不仅仅是场馆空间的延续，还为新博物馆的核心功能提供了空间。

继续向里走，新大楼前是一间老车间建筑，与原博物馆主场馆大厅相连，从而令空间不足的问题迎刃而解。这座混凝土结构建筑的历史可追溯到1987年，总共两层，建筑面积为632平方米，整体经重新刷白，让原本残破的建筑焕然一新。

THE MUSEUM BUILDING

From its West Changjiang Road access point, the core of the Shanghai Museum of Glass extends from south to north. Its architecture consists of five distinct sections: a newly-constructed L-shaped building, a former workshop, atrium, original glass factory, and the redesigned appendix.

The museum architecture retains characteristic features of the original structures – namely the windows – whilst taking advantage of their open-volume spaciousness to efficient effect. By layering these former industrial buildings with contemporary interventions – the facade, atrium and annex – the space's industrial tone is preserved, with its current role as a new kind of museum for China reinforced through modern elements. Dynamic, surprising and offering a clear sequence of spaces through a repeated configuration, the architecture of the Shanghai Museum of Glass speaks to the site's past and present functions, and offers a positive example of the adaptive reuse approach in the city.

The composition embraces the project's wider adaptive reuse model, with new structures integrated where needed to elevate the destination to a Type Two museum and accommodate an extended programme of activities. Featuring a succession of buildings that alternate from new to old, the entire museum is 'framed' by a contemporary glass façade, resulting in high visibility from West Changjiang Road to the south. The striking frontage forms part of the new L-shaped structure that not only extends the venue's floor space but also accommodates core functions of the new museum.

Moving deeper within the site, the new building fronts a former workshop, linked to the original designated museum hall to counter a lack of space. Dating from 1987, the concrete structure spans two floors and 632 sqm. The unit was re-plastered and painted white, considerably brightening the once dilapidated site.

老车间的北面是一块全新的区域：玻璃中庭。它由原先位于两栋建筑间的狭小走道围合而成，呈现一块非常现代的封闭空间。中庭高度提升至老建筑顶端，美轮美奂的质感令其时刻散发着由内而外的独特气息，与两侧老建筑的传统美学氛围形成鲜明的对比。

中庭连接了轴线序列中的第二个老建筑，即原上海轻工玻璃公司旧厂的主要生产车间。它始建于1991年，最高处达20.2米，为带钢窗的混凝土结构，主要由开放式大型空间构成，北面和西面分别有一个副楼，共同构成了其达到2 925平方米的大体量。设计完整保留了原有窗户，外侧刷白。最北面的副楼外立面重新做过处理，旨在展现博物馆前卫现代的一面，同时也起到了在新老建筑间过渡的作用。

Directly north of the former workshop is a new intervention: the glass atrium. The contemporary feature connects the two existing buildings by effectively closing in a narrow alley that once ran between the two. Extending the full height of the original structures, the atrium is an impressive addition from both inside and out, and contrasts starkly with the more traditional aesthetic of the original structures either side in terms of material and modernity.

Next in the sequence and also linked to the atrium is the former main production hall of the factory. Built in 1991 and extending to a height of 20.2 metres, it is a concrete structure with steel-framed windows. Comprising mostly open hall space with two appendices to the north and west respectively, at 2 925 sqm it is one of the largest volumes on the site. Its original windows have been kept intact, and a coat of white plaster applied to its exterior. The structure's northernmost annex was re-clad to reference the contemporary façade of the main museum hall, completing the architectural configuration and sequence of new-old-new.

1　入口大厅
　　Entrance Hall

2　博物馆商店
　　Museum Shop

3　中庭
　　Atrium

4　固定展览
　　Permanent Exhibition

5　灯工工作室
　　Flame Working Performance

6　浇铸玻璃工作室
　　Casting Glass Studio

7　游客中心
　　Visitor Centre

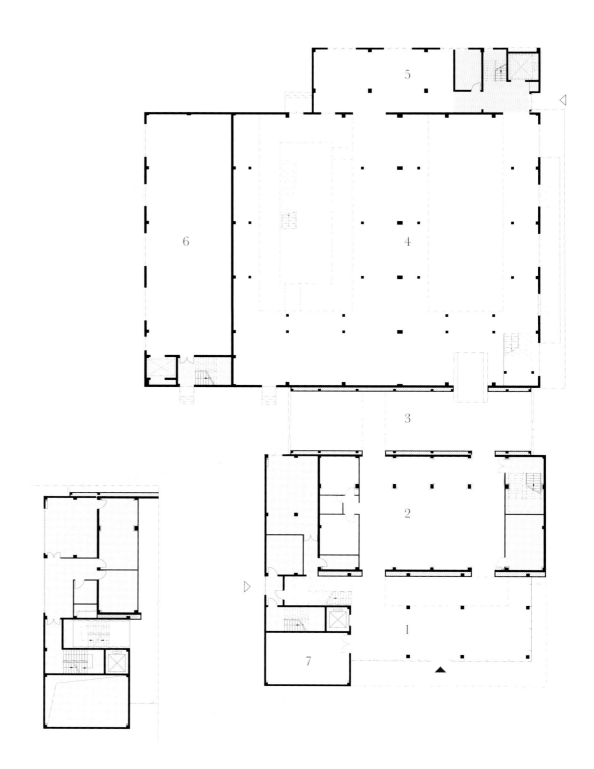

First Floor Plan 1/400
一层平面图

1   固定展览
      Permanent Exhibition

2   临时展览
      Temporary Exhibition

3   博物馆咖啡吧
      Museum Café

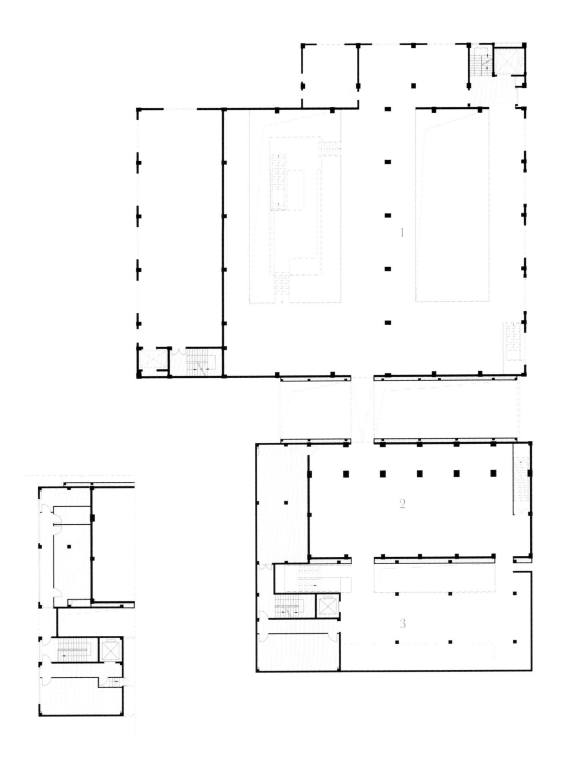

Second Floor Plan 1/400
二层平面图

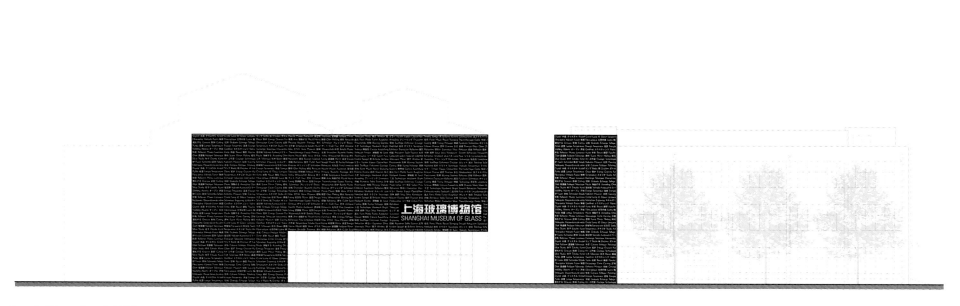

South Elevation 1/400
南立面图

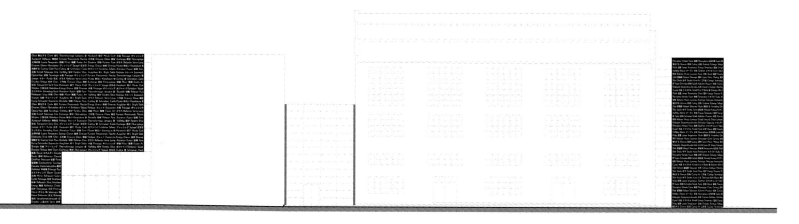

East Elevation 1/400
东立面图

L形结构

设计在老车间建筑周围特别构思了L形钢结构，将南面和西面完全包覆，增加空间的同时还令建筑与道路间的距离缩短了12.5米，特别是留白效果令外立面更加凸显，提升了博物馆品牌的辨识度。站在唯一可进入博物馆的长江西路上，能够总览博物馆全景，这对场馆作为宝山区地标建筑的成败与定位起着至关重要的作用。摆脱其所在地的工业化烙印，进而转化为休闲文化目的地，这是博物馆长远的目标。新建筑让人过目不忘的外立面将"展示玻璃无尽可能性"的终极理念与这一目标联系了起来。

博物馆规划的基本要求是设置能够兼容不同功能和设施的巨大空间，确保达到国际第二类博物馆的标准。这个全新打造的L形结构为博物馆增加了1 198平方米的空间，从而形成了一块功能区域，将博物馆的入口大厅、

接待台、游客中心、员工办公室、礼品商店以及二楼咖啡吧等设施都纳入其中。该结构打破了新旧建筑的既定模式，为博物馆赋予了独具创意的建筑风格：全新的环绕结构通向旧车间，充满现代感的玻璃材质打造的中庭一直延伸到后侧的主展览大厅。

老建筑在布局方面有一定的局限性，而新建的L形建筑则能够灵活分隔，从而满足了小规模、灵活空间的要求。

新建筑巧妙地解决了原有建筑的技术难题，成为连接后侧老建筑的纽带，同时也可以容纳现代博物馆所需的关键基础设施。较之在新建筑中直接进行设备安装，对老建筑的空调、通风管道和水管等设施进行改造，不仅非常费时，而且需要大量资金。因此水、电、煤的线路和管道只从老建筑内进出，而配套设施都安排在新建筑中。

THE L-SHAPED STRUCTURE

An additional purpose-built L-shaped steel structure wraps around the former workshop, fully enveloping both south- and west-facing sides to generate extra floor space. The intervention brought the building an additional 12.5 metres closer to the road, and just as importantly, created a blank canvas that now boasts the striking façade that has gone on to effectively visually brand the museum. Visible from the site's sole access point on West Changjiang Road, these are the venue's principle vistas and as such are crucial to the museum's success and positioning as a beacon for Baoshan district. As a destination museum that aims to set itself apart from a locale better known for industry than culture, such distinction is paramount to its longer-term ambitions. To achieve that goal, a striking façade fronts the new building, linking thematically to the museum's ultimate concept of describing the myriad possibilities of glass.

A fundamental requirement of the planned museum was ample space to house additional functions and facilities to elevate the venue to an international standard, Type Two destination. For that reason, the new L-shaped structure boosts the venue's square footage by an extra 1 198 sqm, resulting in a functional space that now houses the museum's entrance area, reception desk and visitor centre, as well as staff offices, a museum shop and a café on the second level. Kicking off the museum's established pattern of alternating between old and new structures, the building contributes to the venue's unique architectural composition: contemporary wraparound structure, to former workshop, through to modern glass atrium and finally the main exhibition hall behind.

In addition, where the older structures presented physical constraints of layout, the new L-shaped building is more easily divisible, offering opportunities for smaller-scale, flexible spaces.

The new building offers a practical solution to technical challenges posed by the site's original architecture, effectively acting as a service belt to the older structures behind and housing key amenities crucial to a modern museum. Adapting the older building for facilities such as air-conditioning, ventilation shafts, kitchens and plumbing, would have not only proved time-consuming, it would also have been significantly more costly than housing them in the new build. Instead, wires and pipes for water, gas and electricity only travel in and out of the older buildings, with supporting units housed in the more recent construction.

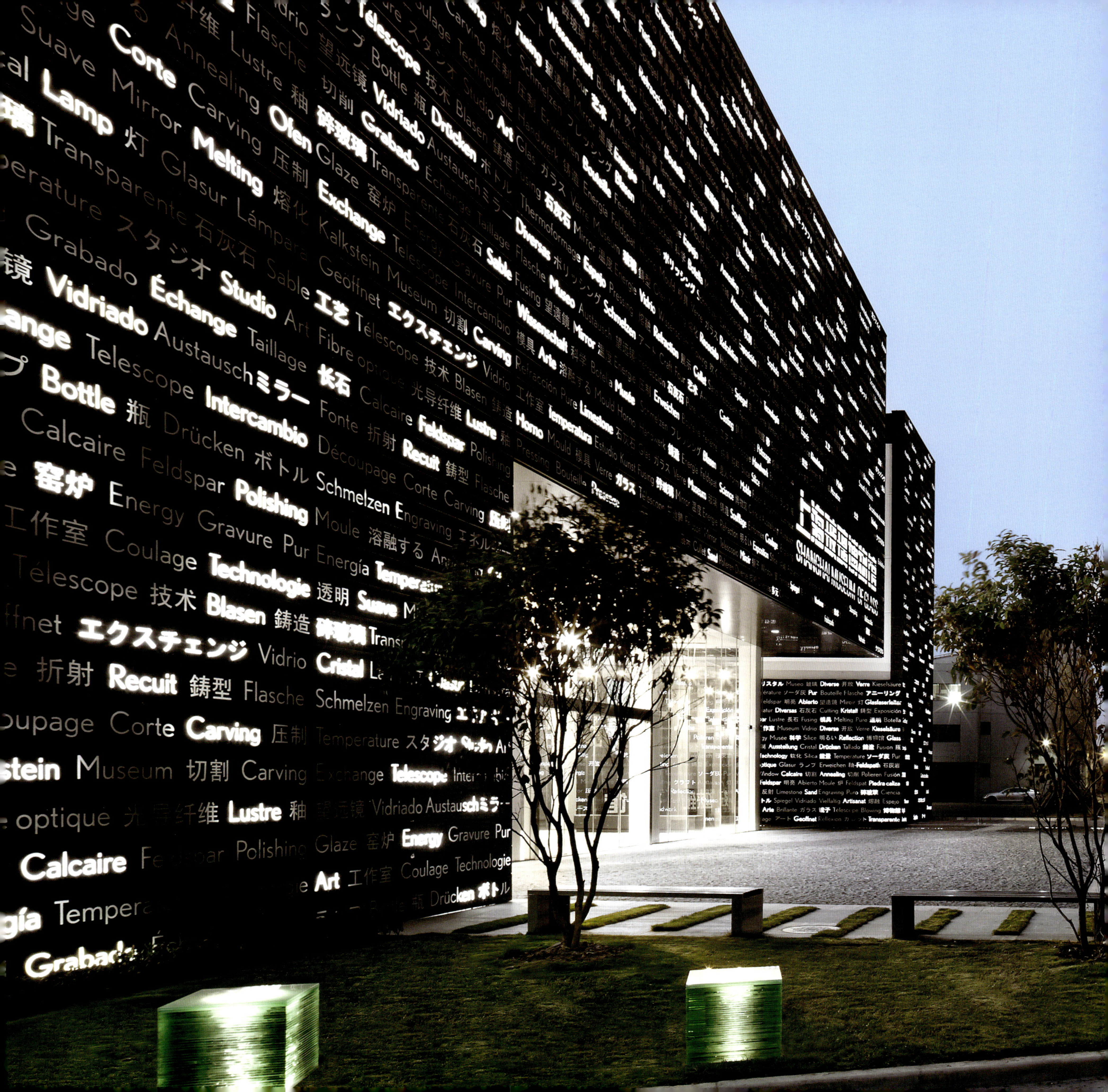
上海玻璃博物馆
SHANGHAI MUSEUM OF GLASS

外立面

博物馆的外立面通过主题理念与材料的应用，在原有建筑与其固有形象间形成联系。外立面是玻璃博物馆建筑的一大核心特色，也已然成为其最具辨识度的元素。惊艳的外立面是该项目成为宝山地标性建筑的关键所在，对这片并不知名的休闲区域来说，高辨识度将在很大程度上促进人气的聚集。

从道路上看去，博物馆是一个身居广场之后的巨大黑色结构。而夜幕降临时人们则能看到完全相反的效果：博物馆正面精心布置的绚丽LED灯让建筑得以时刻保持高可见度。

博物馆南面入口是能够将内部空间完美呈现的透明玻璃。玻璃结构上端的构造浓重而厚实，由大量片状结构组成。每片结构上都用浅色呈现出7种不同语言中与玻璃相关的词汇。

FAÇADE

Linked to the original buildings and their former incarnations via theme and material, the museum's façade is a key feature of the museum's architecture, and has gone on to become a core visual identifier for the project. A striking façade was a key requirement in positioning the destination as a Baoshan beacon, ensuring visibility in an area not known for its leisure possibilities, and helping to overcome the challenges of the museum's remote location.

From the street, the museum appears as a dark, monolithic structure, set deep inside a generous plaza. As night falls, an inverse effect is revealed as strategically positioned LED's light up various points on the museum's arresting frontage, ensuring visibility at all times.

On approaching the museum's south-facing entrance, what appears from afar to be a dense, dark entity above a transparent glass window that offers clear views inside the venue is revealed to comprise scores of tiles. Each panel features a word relating to glass, rendered in a lighter shade, and spanning a total of seven languages.

　　越过广场远眺博物馆，最引人注目的外立面设计由1200多块标准U形工业玻璃板组成。U形工业玻璃板是一种常见的低廉材料，常用于厂房的采光，一般竖向铺设。不过在这里，玻璃板被横向铺设，打造出完全不一样的效果：黑色的涂层和内刻镂空的文字使这种标准材料构建出一种全新的外立面整体形式。

Towering above the plaza, the façade is the museum's most eye-catching feature, and was created from over 1200 standard industry U-shaped glass panels. A common, inexpensive material, they are typically used to bring light into factories and are generally applied vertically. Here, however, the panels are placed horizontally to produce an altogether different effect: with a dark coating and inscribed words, the material is transformed into a new, one of a kind façade.

制作过程

Ⅰ. 准备

U形玻璃是从德国进口，因为本地的供应商所提供的材料无法符合要求。加工的第一步就是清洁材料表面的杂渍。

Ⅱ. 涂层

U形玻璃里层表面镀上黑色涂层。这种涂层是一种轻质防水层，玻璃也因此而变得不透明。

Ⅲ. 强化

涂层需要经过强化处理，在炉内烧制120分钟，这样才能使涂层长久地覆在玻璃上。烧制的温度需要400摄氏度。

Ⅳ. 喷砂

经过喷砂的处理，与玻璃相关的文字才从涂层的表面显现出来，LED背光之后，光线从透明的文字上投射出来。

THE MANUFACTURING PROCESS

Ⅰ. Preparation

The U-shaped channel glass is imported from Germany since local suppliers could not provide material with the required properties. As a first step, all remains must be removed from the material surface.

Ⅱ. Coating

The black enamel coating is applied on the inner side of the channel glass surface. The coating is a light proof layer and the glass now is not transparent anymore.

Ⅲ. Hardening

The coating needs to be hardened for 120 minutes in an oven. This will permanently apply the coating to the channel glass. A temperature of 400 degrees is required.

Ⅳ. Sand-blasting

Sand-blasting is used to reveal glass-related words and characters on the enameled surface. These transparent areas will later be backlit by LED lighting.

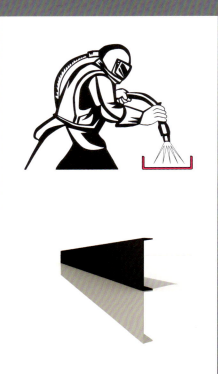

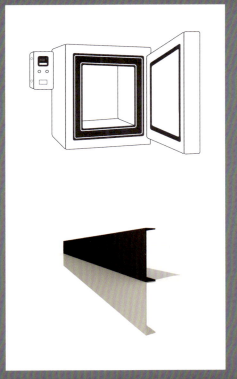

coffnet Telescope 能量 Fonte Ciencia
望遠鏡 Glasfaserleiter Coulage
Calcaire Carbonato sódico 長石
Habilidad 碎玻璃 Échange Four
エクスチェンジ Blasen Ouvert
光滑 Mirror Refracción Calcin
Cullet Polissage 科学 Strahlend
純净 Reflection Silice Scherben
Energy 釉薬 Reflection Cristal
屈折 Découpage 长石 Pressing
Horno ボトル 瓶 Telescopio 釉
Arena Einformen 多元 Brillante
抛光 Le carbonate de soude 灯
Drücken 二氧化硅 Verre 多様

安全至关重要，而在中国很难买到高强度的板材。因此，博物馆决定向德国Bauglasindustrie公司统一采购。该公司在玻璃板材方面的贡献使得博物馆关于工艺与传统的理念也得以加强。板材在欧洲生产后运至中国，接受进一步处理和使用。

运至上海之后，板材经过彻底清理，去除运输过程中出现的所有污迹。每块板材内侧首先经过喷砂处理，随后贴上文字贴纸，再涂上黑色搪瓷涂层，然后在400摄氏度的专业烘箱中接受120分钟的热处理，以令黑色搪瓷涂层稳固。这样文字部分没

有被涂层覆盖，玻璃仍有一定的透明度，通过板材背后精心布置的LED灯便能带来无与伦比的现场效果。

板材上用7种语言展现出与玻璃相关的文字，与博物馆的内涵与主题浑然天成。东西方各种语言间的平衡，强化了上海玻璃博物馆的国际性，尤其是其布展方式、社区概念以及中国顶级文化机构的品牌定位。标新立异地在外立面上使用标准U形玻璃板，博物馆开创性地实现了自己的首要目标：以引人入胜、独具匠心而又魅力四射的方式来展示玻璃这种材料。

Safety was paramount, and it proved impossible to source panels of sufficient quality in China. Instead, the panels were manufactured in Germany by Bauglasindustrie GmbH, whose contribution to the museum reinforces wider concepts of craftsmanship and tradition. Produced in Europe, the tiles were then transported to China for further treatment and application.

Upon arrival in Shanghai, the panels were thoroughly cleaned to remove all contaminating traces left by handling and transportation. A black enamel coating was applied to the inside of each tile, before being heat-treated in a specialized oven at a temperature of 400 degrees for 120 minutes. The procedure effectively 'fixed' the black coating, rendering it resistant to wear and tear. The final step in the process was to sandblast textual inscriptions into each panel, revealing the glass' inherent transparency behind, further enhanced by strategically placed LED lights.

Words in seven different languages relating to glass link the museum's outward appearance to its content and theme. This balance of languages from both east and west reinforces the international intentions of the Shanghai Museum of Glass in terms of its exhibits, concept of community, and positioning as a globally-significant institution for China. Through the creative application of standard U-shaped glass tiles, the façade further underpins the overarching purpose of the museum: to present glass in ways that are surprising, inspiring and engaging.

ique 光导纤维 **Lustre** 釉 Pressing Bouteille **Pressage** 吹く 熔融

alcaire **Feldspar** Polishing

折射 **Recuit** 鋳型 Flasche

age Corte Carving **压制**

en Engraving エネルギー

熔融する Annealing **Ofen**

Temperatur 温度 Energie

ave Mirror **Melting** 熔化

ransparente 石灰石 Sable

amp 灯 **Glasur** Lámpara

en Engraving **エネルギー**

ture **スタジオ Studio** Art

e **Telescope** Intercambio

Vidriado Austausch**ミラー**

焊炉 **Energy** Gravure Pur

室 Coulage Technologie

Bottle 瓶 **Drücken** ボトル

入口

进入博物馆的参观者在到达主展示区前需经过几块区域。博物馆在其中利用光影、新旧、开放与封闭等一系列环境设计的对比，既为观众的参观之旅做好了铺垫，同时也丰富了他们的体验与感受。40米进深的广场是游览的第一站，广阔的空间营造出庄重而大气的氛围，这在寸土寸金的城市中实属难能可贵。大部分参观者都会先在这里稍作停留，再分批步入博物馆。

博物馆的主入口位于整体建筑上内凹的区域，底层几乎都使用了轻质透明玻璃。入口大厅高达6.6米，令人印象深刻。这一区域的装饰风格优雅摩登，与博物馆本身的内涵与用途相得益彰，让人们对即将观赏到的高品质藏品充满期待。入口的开放式结构受工业厂房的历史启发而打造，大厅则采用了现代感十足的设计理念。

ENTRANCE AREA

On entering the museum area, visitors pass through a sequence of spaces before entering the main exhibition, designed to build anticipation and to enrich visitors' experience through contrasting environments of light and dark, new and old, open and closed. First is the 40-metre-deep plaza. Not only does it serve to set the scene for the museum's architecture, in a city where space is at a premium, the plaza is a particularly generous proposition, imbuing the setting with an air of formality. Visitors can pause here for a moment and gather in small groups before entering the museum.

The main entrance is situated in a recessed area at the foot of the museum's frontage. A light and transparent glass front spans almost the entirety of the museum's ground floor and opens onto the 6.6-metre-high entrance lobby. A grand and dignified space, it is in keeping with a museum of this scale and ambition, and indicates to visitors the quality of exhibits ahead. While the exposed structural elements of the entrance hall reference the industrial past of the site, the overall design of the lobby is modern.

　　接下来是一段没有窗户、高度保持一致的200平方米开阔空间，用作博物馆商店。这一空间是老车间的一部分，如今与博物馆建筑巧妙地融入在一起。参观者从此处进入阳光充足的中庭，接触到另一种与众不同的环境，从而在博物馆的现代特质中获得更深刻的体验。设于建筑二层悬空过道的老车间墙体高度在改造中得到提升，无不展现着空间的开敞性。明亮的中庭是进入主展区前最后一块可供停留的区域，紧接着便是通向玻璃博物馆主展馆的幽暗镜面走廊。

Next in the succession is a windowless, spacious room, similar in height and offering 200 sqm, now used as a museum shop. Part of the former workshop, it has been integrated seamlessly into the museum complex. From here, visitors step into the sunlit atrium space, offering yet another distinctive environment to enhance experience and reinforce the contemporary focus of the museum. Extending the entire height of the former workshop, its sense of openness is furthered through suspended bridges on the building's second floor. The last stop before entering the main exhibition area, the contrast between exhibition spaces and atrium could not be more stark: from a bright, airy environment, visitors step into a dark, mirrored corridor leading into the museum main hall.

博物馆策划 MUSEUM PLANNING

博物馆体验

上海玻璃博物馆选择"第二类博物馆"开发模式，这一定位在观众的博物馆体验环节得到共鸣和升华。中国与日俱增的文化机构与设施大多以单一直接的陈列或"历史庙堂式"的布局为主。上海玻璃博物馆为凸显自身特色，致力于建成中国第一家真正意义上的现代动感博物馆。

为此，博物馆的策略强调互动的叙事方式。观众可以把整个博物馆当作一个主题故事来解读，其内容涵盖了玻璃材料的科学、历史、应用和创新。这种非线性方法可从多个层面上吸引参观者：对于发烧友级爱好者，这种方式可以令其从头到尾完整地进行参观；而普通观众则可自主决定融入"故事"，或是全身抽离出来，根据自身喜好欣赏展出作品。一切都围绕着现场的核心氛围，即"分享玻璃所带来的无限可能"。

展品布局与博物馆策划相得益彰，各展示单元围合成圆形，其间设有咖啡厅、休息位等舒适区域。藏品展示始终融合各种形式的观众互动，现场演示则衍生出材料的发展史及现代应用，更有一处关于现代玻璃设计的专区令过去与当下遥相呼应。

博物馆推出旨在促进交流、娱乐和教育的公众活动计划，为其各个空间带来愉悦参与氛围的同时，亦培养出观众的认同感。此类活动的广阔空间是博物馆早在设计伊始便充分考虑其中的。

博物馆体验紧扣"链接过去、当下与未来"的宗旨。位于历史悠久的上海轻工玻璃厂旧址，这本身便呼应了展览的叙事题材，传达出将历史引入生活的理念，力求满怀激情地重燃传统工艺的光辉。

MUSEUM EXPERIENCE

The Type Two model on which the positioning for the Shanghai Museum of Glass is based is further crystallized in the museum experience. To counter the straightforward showcases or 'temples of history' that characterize China's increasingly saturated cultural landscape, the Shanghai Museum of Glass aims to be China's first 'living' museum.

A central strategy in this regard is an emphasis on interactive storytelling. The museum can effectively be 'read' as a thematic story, charting the science, history, applications, and creative possibilities of the material. This non-linear approach works to engage visitors on several levels: for the dedicated enthusiast, it can be followed closely from start to finish. Just as meaningful, visitors can effectively opt in and out of the museum's narrative and engage with works on display on their own terms, without detracting from the venue's core message of sharing the countless possibilities of glass.

Exhibits' physical layout lends itself to this planning strategy, with units presented in a circular route, punctuated by convenient resting points such as the museum's café Murano@SHMOG and seating areas. Throughout, displays incorporate various means of engagement, with live demonstrations effectively linking the material's history with its contemporary applications.

A program of museum events focused on exchange, entertainment and education brings liveliness to the museum's various spaces and creates a sense of community. In planning the layout for the Shanghai Museum of Glass, square metres were intentionally sacrificed to create room for activities of this kind within the museum environment.

The museum experience revolves around the central concept of linking past, present and future. A goal in part defined by the venue's rich history as the former premises of Shanghai Glass Company Ltd., it also ties in closely with the storytelling element of the museum, bringing history to life, and reigniting traditional crafts in an exciting, contemporary way.

目标群体

按照"第二类博物馆"模式,上海玻璃博物馆必须使现场突破简单展示的惯例。博物馆是集众多功能于一身的场所,允许不同类型的观众在不同的时间段进入其中参与体验,特别应该对以家庭为单位的观众给予更多关注。随着中国的迅速发展,人们合家共享的闲暇时光已弥足珍贵。中国传统意义上的休闲时间一般用于娱乐、社交和学习,能够同时满足上述需求,又可以提供整套服务的文化活动便是中国家庭共享天伦的理想选择。为此,上海玻璃博物馆正努力打造能够使三代同堂、共同学习、体验新鲜事物并参与趣味活动的多功能空间。

TARGET GROUPS

Following the Type Two model, the Shanghai Museum of Glass' experience is planned in a way to elevate the venue beyond a straightforward showcase. A multipurpose destination, it encourages multiple-hour visits for a broad range of target groups, with a strong focus on families. In a rapidly developing China, leisure time spent with family has become increasingly scarce, and thus more important and by extension, valuable. Traditionally, leisure time in China centres on relaxation, social interaction and learning. Families seek activities that combine these factors, and that offer packaged services to compensate for the lack of time available for spending time together. In creating an environment where grandparents, parents and children can learn together, enjoy new experiences, and partake in fun activities, the Shanghai Museum of Glass underlines its commitment to connecting with its community.

　　年轻人也是上海玻璃博物馆的主要参观群体。出于接受高等教育及规划职业生涯等原因，越来越多生活在城市的中国年轻人选择将结婚生子等人生大事延后，于是逐步形成了一个拥有足够时间进行休闲娱乐、外出旅行或与家人朋友共享假日时光的中产阶层。夜店酒吧、餐饮和购物中心等商业场所已开始将目标瞄准了这一群体。而作为众多休闲选择之一，博物馆可通过寓教于乐的方式吸引这一阶层，利用网络对年轻人进行行为观念的影响，从而提升自身的品牌及知名度。

Besides families with young children, a key demographic for the Shanghai Museum of Glass are young influencers. As young city-dwelling Chinese increasingly delay life events like marriage and having children in favour of education and career, an affluent group has emerged with more time for recreation, entertainment and travel, as well as spending time with family and friends. Already, myriad services, venues and trends vie for their attention and time, including bars, clubs, restaurants and shopping malls. In offering an alternative, culture-driven source of entertainment, the museum engages with this group, while benefiting from their prominent role in influencing trends amongst young Chinese adults online, helping shape the reputation of the Shanghai Museum of Glass.

上海玻璃博物馆将自身定位为中国首开先河、独具现代特色的博物馆，借此吸引国际游客。在上海逗留期间参观博物馆本身就是对艺术和设计的一种认同，这些国际游客的到来对提升博物馆的全球声誉颇有裨益。为满足他们的需求，上海玻璃博物馆在投入运营后第二年便引入了三语服务。

Through its positioning as a modern design destination in Shanghai, and altogether new kind of venue for China, the museum pitches to an international group of travellers and tourists. Foreign tourists who visit the Shanghai Museum of Glass during their stays in the city have an established interest in art and design, and are crucial to building the museum's global reputation. To further cater to this group, the Shanghai Museum of Glass introduced tri-lingual communication in its second operational year.

观光之旅

　　上海玻璃博物馆通过一个个精彩纷呈的展览单元对玻璃进行了全面介绍，观众可以自由选择全面参观或是单独鉴赏。博物馆游览行程一分为二，第一部分为首层游览。首层可归类为知识教育区，提供了有关玻璃特性的众多信息、玻璃在中国和西方的发展历史以及玻璃令人叹为观止的各种用途。

　　万花筒般的绚丽入口给观众留下了博物馆现代设计的第一印象，它完美展现着玻璃的技术性和艺术可塑性，为进一步观光奠定了良好的基调。从这里开始，第一单元的主题为"什么是玻璃"，向观众介绍玻璃这种材料的核心本质和化学结构。通过亲眼所见及触手可及的体验，观众从一开始便掌握了玻璃的基本知识。

　　第二单元主题为"技术和工艺的发展"，阐述的是玻璃生产在中国和西方各自的发展路径与轨迹。本单元旨在介绍玻璃这种材料在两种文化背景中的悠久历史，使观众能够近距离观赏稀有罕见的古代玻璃珍品。同博物馆其他区域一样，这样的展示可避免向观众单向灌输史实的套路。博物馆把观众需要的信息有所区别地呈现出来，让他们主动展开各自不同的体验。

VISITOR JOURNEY

All units of the Shanghai Museum of Glass fit together to tell an overall narrative of glass, but can equally be enjoyed separately. The flow of visitors around the space leads museumgoers on a two-part journey, the first of which is situated on the ground floor. The space can be categorized as educational, and includes information on the properties of glass, its history in China and the West, and its surprising applications.

The 'Kaleidoscopic Entrance' sets the tone for the first part of the visitor journey by creating a positive first impression of the contemporary nature of the venue, its commitment to technology, and emphasis on the artistic possibilities of glass. From there, a first unit titled 'What is Glass' aims to introduce visitors to the core qualities of the material, as well as its chemical structure. By touching on hard facts early on, visitors are equipped with a basic knowledge of glass from the outset.

A second unit devoted to 'Development of Technology and Craftsmanship' describes the legacy of glass production in China and the West respectively. The aim of this unit is to celebrate and illustrate the material's rich history in both cultures, and to bring visitors face-to-face with rare ancient artifacts. As with other areas of the venue, the strategy avoids overwhelming visitors with facts. Instead, visitors are invited to explore the collection at their desired level of detail by offering information in a layered manner, putting them in control of their own museum experience.

　　展示工坊向观众现场演示传统技术从过去到现在的演变过程。这种引人入胜的感官体验与之前强调事实陈述的单元相互穿插，再次抓住了观众的眼球。首层第三个单元，同时也是最后一个单元，以"从日常生活到科技前沿"为主题，逐一介绍玻璃在当今社会令人惊叹的应用，涵盖范围从日常家用到尖端科技，无所不包。这个单元唤起了观众的个体记忆与经验，生动有趣而又平易近人，体现出玻璃与日常生活密不可分的重要联系。

An integrated workshop, where traditional glass techniques are presented live for the visitor to see, forms a transition back from past to present. This engaging, sense-driven feature punctuates the preceding fact-heavy units, recapturing visitors' attention. The final and third unit on ground floor level, 'From Daily Life to Cutting-Edge Technology', invites visitors to literally uncover surprising present-day applications of glass, ranging from the domestic through to the high-tech. The section relates to visitors' personal memories and experiences of the material, connecting in a fun, accessible way and reminding museum-goers of glass' ubiquity and significance in daily life.

　　楼梯通往二楼夹层，展示的内容也在这里由博物馆教育过渡为侧重玻璃的艺术可塑性。同首层的寓教于乐和动手体验相比，二楼拥有着尊贵专享的基调。静谧的环境更像是一个当代艺术画廊，观众在这里回望与沉思首层的展示，从而更深入地理解玻璃的历史、科技和应用。与前几个单元更强调玻璃应用的普遍性不同，二楼想要展现的是玻璃的美丽与神秘——玻璃本身就是一种艺术形式。一处狭小而富有灵性的艺术空间强化了这里高贵奢华的氛围，在恰到好处的私密环境中，无声似有声地展示着艺术珍品。

A staircase leading up to the second floor mezzanine level marks a juncture between the museum's educational elements, to an area dedicated to more artistic applications of glass. Contrasting with the first floor's focus on hands-on, fun elements, the second floor sets a tone of exclusivity. A hushed environment more akin to a contemporary art gallery, its positioning invites reflection, with visitors able to look back and down on the first floor's exhibits of history, science and applications of glass. Contrasting with earlier units' emphasis on the ubiquity of glass, the second floor has been planned to allude to its beauty and mystery that here serve to elevate the material to an art form in itself. Furthering the tone of exclusivity is a small spot-lit gallery space used to display miniature, precious objects in an appropriately intimate environment.

上海玻璃博物馆的参观范围绝不仅限于主馆，其他空间的植入和设计丰富并强化了观众在主馆所获取的信息与体验。热玻璃表演厅和DIY创意工坊位于博物馆主厅之外，直观的演示栩栩如生地呈现出了玻璃奇妙、丰富而又无穷无尽的创作潜力。对老厂房的参观专注于玻璃材料在变化发展中的实用性及再生产沿革；而Keep it Glassy国际创意玻璃设计展则邀请国际范围内的艺术家通过各自的创作挖掘玻璃的无限艺术潜能。这里既是博物馆主厅叙事清晰的延伸，同时也可以作为独立的艺术展单独鉴赏。

The visitor journey extends beyond the main museum hall, with additional elements designed to reinforce key messages. The Hot Glass Performance Hall and DIY Creative Workshop both take glass outside of a museum setting, bringing it to life through active, highly visual demonstrations that reinforce the underlying magic, versatility and possibilities of glass. A factory tour brings the focus back to the material's ongoing relevance and continued production, whilst Keep it Glassy, an exhibition of contemporary design pieces by international artists, underlines the endless creative potential of glass. The additional functions are a clear extension of the story told in the museum main hall, but yet again can be appreciated and enjoyed separately.

成长模式

上海玻璃博物馆追求不同层次的内容策略, 贴近尽可能广泛的目标群体, 同时注重培养博物馆自身的社交群落。展望未来, 博物馆各项新增功能将有的放矢地吸引特定受众, 并最终促进博物馆理念的持续强化和提升。

除举办定期展览之外, 博物馆还推出了一项艺术家驻馆计划, 为热衷从事玻璃创作的艺术家提供设施齐全的住所与工作室。此项计划可以为博物馆带来专业技术作品、支持艺术创作, 并让观众得以一睹专业玻璃艺术家的创作实践过程。江西籍艺术家王沁现任上海大学美术学院玻璃工作室教师, 他是该计划的首名入驻者。

当代玻璃艺术空间伽蓝画廊 (Glaze Gallery) 对艺术家们来说是另一个千载难逢的创作机会, 因为他们可以在专享的艺术创作环境中展示自己的作品。这一举措使博物馆得以吸引业内颇有分量的藏家和行家, 他们的支持是推进博物馆走向成功的巨大动力。

珂庐 (KILN) 是一家私人会所, 在博物馆的跨界文化与商业关系领域贡献颇丰。会所可为从小规模展览到大型宴请等各类活动提供尊享的私密空间。对于希望在考究环境中宴请宾朋的博物馆VIP来讲, 会所雅致细腻的设计与环境无疑是他们的最佳选择。此外, 珂庐还可为博物馆正在酝酿和开发中的婚礼中心提供副场地。

GROWING MODEL

The Shanghai Museum of Glass pursues a strategy of layering content, with services and approaches to appeal to a broad range of visitors, all the while building a museum community. Looking forward, new functions will help the museum to reach out to specific new audiences, ultimately contributing to a sustainable concept.

In addition to a museum building that can host temporary exhibitions showing upcoming as well as established artists, an artist-in-residence scheme offers on-site accommodation and a fully-equipped studio space to artists actively engaged in the field of glass production. The programme exists not only to bring specific skills and expertise to the museum and support the arts, but also to offer visitors a glimpse into the practice of professional glass artists. Jiang Xi-born talent Wang Qin, a working artist and instructor at the Glass Studio of the Fine Arts College of Shanghai University, was the first to participate.

Specialist contemporary art space, Glaze Gallery, creates additional opportunities for artists to showcase their work in an exclusive, art-oriented environment. It helps the museum to attract an important community of collectors and connoisseurs whose support is vital to the museum's success.

Aimed at cultivating relationships in both culture and business is KILN, a members-only club. For the museum, the venue offers an exclusive space for events of all kinds, from small-scale exhibitions to banquets. Its elegant setting offers a retreat to VIPs seeking a sophisticated destination for entertaining guests. Equally important, KILN will serve as a secondary location to a wedding venue that is set to be developed on the museum's premises.

内容开发

有别于传统博物馆，收藏并非上海玻璃博物馆的建馆初衷。计划展出的藏品主要为博物馆所有者和经营者们的私人珍藏，这种做法固然不错，但不足以为广泛的参观者提供耳目一新的博物馆体验。此外在中国当前的文化机构中，艺术品的国际采购环节仍然困难重重，而且成本颇高。因此，进行场馆布展的同时，上海玻璃博物馆还应并行不悖地采取其他策略。

博物馆间互借、向艺术家和藏家直购并保持密切合作等方式当然能够促使博物馆藏品与日俱增，但内容的开发还可以拥有更为创新的方式——向中国本土艺术家和设计师购买全新艺术作品。以俄国作家契诃夫的诗句为造型的艺术荧光灯管、别具一格的彩绘玻璃窗以及灵感源自珍妮·霍尔泽 (Jenny Holzer) 创作的滚动LED屏等作品既融入了博物馆设计本身，又同为玻璃创作的藏品。正在开展的艺术家驻馆计划在推动博物馆收藏发展的同时，又可培育艺术新人，可谓一石二鸟。更令人欣喜的是，日常生活物件和现场表演都已引入到了博物馆体验当中，成为博物馆叙事的有机组成部分。这种与观众对玻璃材料的个人体验息息相关的展示，通过趣味盎然而又平易近人的方式呈现了玻璃的当下与未来。

CONTENT DEVELOPMENT

Different from traditional museums, the Shanghai Museum of Glass is built around a theme, as opposed to on a collection. Initially, the objects to be showcased in the museum comprised predominantly the personal collection of the museum's owner and director. Although appealing, they were not sufficient to create an innovative museum experience for a broad audience of visitors. Additionally, international purchasing of art objects remains challenging and costly in China's cultural landscape. Therefore, parallel to the planning of the venue, different strategies were chosen to create content for the Shanghai Museum of Glass.

Naturally, the museum's collection has been gradually extended through inter-museum loans, direct purchases and close collaboration with artists and collectors. But more innovative means of content development include the commissioning of new artworks by China-based artists and designers. A quote by Russian author Anton Chekhov in fluorescent tube lighting, unique stained glass windows, and a Jenny Holzer-inspired rolling LED screen double as design elements as well as collection pieces. An ongoing artist-in-residence scheme has a similar dual function, both in furthering the museum's collection and nurturing new talent. More uniquely, daily life objects and live performances have been integrated into the museum experience as part of its story. They relate to visitors' personal experiences of the material, connecting in a fun, accessible way to the present and future of the material.

博物馆设计MUSEUM DESIGN

概念设计

上海玻璃博物馆设计的核心是"连接玻璃的历史、当下与未来"。在集合玻璃及其无限潜能的基础上，设计使博物馆的叙事方式与玻璃的潜能和创意相互交融，从而催生出全新的理念。所有创新元素全部集中于一个现代空间当中，探索并展示着与玻璃相关的万千故事。

博物馆通过业主与管理者、叙事主题及适应性重建的建筑特性，与其前身（即上海轻工玻璃公司的一个生产基地）的功能保持着紧密的历史关联。这一理念在博物馆内得到延续：完整保留了玻璃工厂原有的细节结构和建筑原貌，一系列历史悠久的藏品也展现了玻璃的历史沿革。为强调与21世纪的契合度，保持时代感且永不过时，博物馆采用极具个性的设计以讲述玻璃材料的发展历程及其长久不衰的吸引力所在。

设计还通过互动展品诠释玻璃在当今社会的应用，体现了科技进步对玻璃材料的影响以及玻璃在日常生活中的作用。以玻璃作为起点，博物馆带给观众的是惊喜、启发和联想。无论观众此前是否具备玻璃材料方面的相关知识，为他们中的所有人提供寓教于乐的学习体验是博物馆最大的功能之一。

上海玻璃博物馆不仅承担着简单的物件展示功能，更打造着学习与分享体验的社区空间。博物馆还将引入艺术家驻馆计划和综合活动空间等各种着眼于未来的项目，将内容的前瞻性、丰富性及商业可行性全面纳入考虑。

CONCEPT DESIGN

The link between the past, present and future of glass lies at the heart of the design of the Shanghai Museum of Glass. Taking glass and its seemingly limitless potential as a foundation, narratives, possibilities and ideas are intertwined in ways to give rise to new ones, all housed in a contemporary space that explores and visualizes myriad stories related to glass.

Through its owner and director, theme, and the adaptive reuse nature of its architecture, the museum maintains historical ties with its former function as one of the manufacturing bases of Shanghai Glass Company Ltd. This concept is continued inside the museum venue, where original details are deliberately left intact in order to preserve the buildings' past, and where a collection of ancient objects highlights the history of glass. To create a museum that is relevant in the 21st century and will remain so for the future, new design strategies are used to tell the tale of the material's journey and continued appeal.

Contemporary applications of glass are shown in interactive exhibits, explaining the scientific advances the material has enabled, as well as the role of glass in daily life. In this sense, the museum takes glass as a starting point to surprise, inspire and relate to its visitors. A strong emphasis is placed on creating an educational experience that is both accessible and enjoyable for all, regardless of the viewers' prior knowledge of the material.

By taking its main content as a starting point, the Shanghai Museum of Glass fulfills various roles ranging from the expected – the showcasing of objects – through to the more complex: offering a community space for learning and shared experiences. Incorporating elements that contribute to the future – such as an artist-in-residence scheme and integrated event spaces – the museum venue can be described as forward-looking, multifaceted and commercially viable.

一个万花筒式的入口完成了博物馆公共区域到展厅的过渡。延伸至主要展览空间的通道两侧装有与天花板齐高的折边不锈钢镜面，不同角度的折面反射出视频影片，隐喻着玻璃的内在特质：自成影像，兼具反光功能。如此创意十足的设计为这座由老厂房改造而来的展厅奠定了充满现代感的基调和氛围。

The transition from the museum's public spaces into its exhibition halls is clearly marked by a 'Kaleidoscopic Entrance'. A gateway into the main exhibition space, it comprises ceiling-height video screens whose images are reflected in angled mirrors. Meanwhile, the installation alludes to some of the qualities of glass explored within: both its use in film, and inherent light-reflecting property. Above all, it sets the tone for the contemporary atmosphere of the exhibition halls inside the former factory building.

首层:
学习,发现,参与

上海玻璃博物馆主展览厅反映的基本理念是教育、参与、人才培养,以及历史、当下与未来的衔接。新设计打破了传统惯例,有关玻璃的无数故事被转译为一系列可视作不同"章节"的空间,它们不按时间顺序排列,自成一体,观众可选择单独进行鉴赏。

主展览空间的一楼首先呈现的是玻璃的核心特性,进而介绍玻璃在中国和西方文明中的发展历史。平行柱相向排列,中间隔以炫彩的玻璃地板。光面齐腰的陈列柜展示着中国的古老文物,文字、图片和地图说明令观众的体验之旅更为生动流畅。iPad代表着现代的互动方式,可帮助观众深入理解博物馆主题。颇有深意的灯光和隔断设计彰显出这些珍贵文物的重要价值,而柔和的光线则烘托出亲密与共通的氛围。

LEVEL 1:
LEARNING,
DISCOVERING,
ENGAGING

The main exhibition halls of Shanghai Museum of Glass reflect its underlying concept: education, participation, the nurturing of new talent, and the linking of past, present and future. Going against a traditional set-up, the myriad stories of glass are translated into a sequence of spaces. Equally, they can be considered as 'chapters' to be viewed and enjoyed independently, and which are not chronological.

On the first floor of the main exhibition space, visitors are guided through some of glass' core qualities, before examining the history of its development in China and western civilizations. The parallel accounts stand opposite from one another, separated by a striking glass floor. Glossy, waist-height display cases holding objects from ancient China are supplemented by explanatory texts, images and maps, set alongside iPads to inject a contemporary, interactive means for visitors to delve further into the subject. Careful lighting and segregated displays serve to accentuate the importance of these rare pieces, with the space's low-lighting creating a sense of intimacy and connection.

跨过玻璃地板，从地面延伸至天花板的彩色展示柜呈现在眼前，真实再现着时代较近的西方历史文物，打造出完全不同的视觉效果。精心选编的图像、文本和大事记，从视觉上强化了东西方在玻璃制作方法上的差异，而耀眼夺目的玻璃地板实现了两种叙事之间的衔接。

适用性和功能性是整个博物馆主厅的主旨所在，供临时展览的综合区、小型聚会区和投影播放区点缀其间。这种灵活的多功能设计使博物馆形成了极具魅力的活动空间，也创造了门票销售之外的其他收入来源。

Across a striking glass floor, colourful floor-to-ceiling showcases quite literally offer a different view of later historical objects from the West. Enhanced by carefully selected imagery, text and timelines, the section visually reinforces the difference in approach between East and West, with the eye-catching transparent glass floor providing a thematic link between the two strands of narrative.

Adaptability and functionality are concepts demonstrated across the museum main hall, with integrated areas for temporary staging, small gatherings and projector screens dotted throughout. Such versatility renders the museum an attractive setting for events, securing an additional stream of revenue separate from ticket sales.

观众的整个体验过程都采用数字化方法进行，充分体现了博物馆寓教于乐而又动感现代的特色。互动触摸屏以问答形式揭示玻璃的特性，iPad等现代工具的演示为观众提供了更为具象化的图片和信息。通过透明的玻璃地板，制作玻璃的原材料在脚下清晰可见，一个大型互动装置通过数字游戏来揭示玻璃加热的熔点，趣味十足，令人赞叹。

博物馆以颇具创意的方式演示复杂材料的属性，无需借助繁冗的文字介绍，而是利用数字技术直接揭示事实 (即材料科学和历史探索单元)，这对年轻观众来说更简单易懂。互动环节引导观众灵活自由地探索神秘的玻璃世界，还为不同知识水平的观众提供了对应的参观环境。

Reinforcing the museum's modern style, digital applications are used throughout to educate and engage. Interactive touch screens take a question and answer approach to reveal the core qualities of glass, whilst other displays incorporate iPads that give access to further information and images of objects on show. A large interactive installation allows visitors to heat up glass in a digital game to discover its melting point, while standing on a transparent floor that reveals glass' raw materials.

Explaining hard-to-demonstrate, complex attributes of the material in an accessible way, the need for extensive text is thus negated, with digital technology used to make fact-heavy areas – namely units exploring the science and history of the material – more approachable and relevant, particularly to a younger audience. The interactive solutions allow visitors to explore the subject of glass at their own pace, and create a layered environment that is accessible for visitors with differing levels of knowledge.

东西交流的驿站

ANCIENT PER...
THE MEETING ...
OF THE EAST A...

古代东西方玻璃制造技术交流...

在公元前5世...

Because of its geographical location, in approximately ... ancient Persia learned glassmaking techniques from the Mesopotamian region. By the time the last Persian dynasty (226 -651), the Sassanid Dynasty was established, Persian glass cutting, grinding and engraving techniques had achieved a pinnacle.

The glassware produced in this region has ... very rich ... al flavor, yet regardless of whether one considers it ... the standpoint of technique or style, ... ed Islamic glass. After the first Persian Empire ... ished following the 5th Century B.C. this ... ... mixing point for glassmaking techniques ... east and west.

谜你玻璃瓶
伊斯兰（8-9世纪）
收制，车刻工艺
藏品来自Y&X COLLECTION

Miniature Pot
Islamic, 8th - 9th century
Blown, wheel cut
Collection of the Y&X COLLECTION

蜻蜓眼玻...
伊朗（前300-前20...
直径：9mm
藏品来自Y&X COLLECTION

Dragonfly Eye Glass Beads
Iran (Persia), Pre-Achaemenid to ...
Period, 300-200 B.C.
D. 9mm
...tion of the Y&X COLLECTION

离产生了深远的影响。而且

之后，该地区就一直是

转站

Bottle
Mesopotamia of Iran (Sasanian),
4th-7th century A.D.

Pear shaped body, transparent, pale yellowish
brown glass; blown, wheel-cut, ground and
polished; the top of the bottle is round, while
its base is covered with 16 small convex
horizontal bands

H. 202 mm
Collection of The Corning Museum of Glass,
Corning, NY, USA

玻璃珠
伊斯兰时期

共83颗，（最大）直径：15mm
藏品来自Y&X COLLECTION

Iridescent Glass Beads
Islamic, 7th - 9th century
83 pieces L. (Max) 15mm
Collection of the Y&X COLLECTION

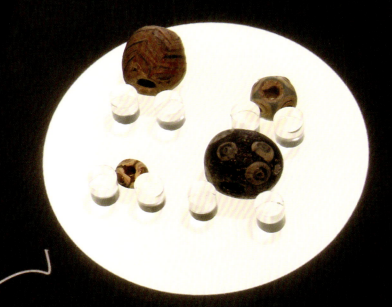

A:

乳棒融合术

兰（7~世纪）
COLLECTION

LACE

蜻蜓眼玻璃珠
中亚地区（公元前5世纪）

直径：12mm
藏品来自Y&X COLLECTION

D THE WEST
th - 8th century
ECTION

Dragon Eye Glass Beads
Middle Asia 00 B.C.
D. 12mm
Collection of the Y&X COLLECTION

珠

马赛克玻璃珠
早期伊斯兰（7-8世纪）

直径：21mm
藏品来自Y&X COLLECTION

Mosaic Glass Bead
Early Islamic , about 7th - 8th century
D. 21mm
Collection of the Y&X COLLECTION

碗
近东地
非洲东
斯（约10

透明的翠
工艺主要为

高：76mm.
现藏于美国康宁

高脚杯
萨珊王朝（约3世纪）

浅绿色玻璃，表面带有浅绿色和深绿色痕
迹。工艺主要为吹制和表面装饰。

高：284mm
现藏于美国康宁玻璃博物馆

介绍玻璃与日常生活紧密联系的单元将观众的注意力带回到现代，这里展示了玻璃从家用器具到尖端技术的各种用途。博物馆邀请瑞士插画家雅克 (Jacques) 和布里奇特 (Brigitte) 创作了一面巨幅插画墙，鼓励观众互动，通过打开插画墙内的暗门来亲手找寻与玻璃应用相关的问题与答案。暗门按儿童身高设计，鼓励幼龄观众独立发现玻璃的众多用途。

A unit focused on the role of glass in daily life brings the focus back to the present, uncovering various applications of glass from domestic through to cutting edge technologies. An illustrated wall, created in collaboration with Swiss illustrators Jacques and Brigitte, stimulates visitors to discover the answers to questions about glass in a hands-on way through the opening of hidden doors in an interactive wall. Positioned at child-height, it prompts younger visitors to uncover the many applications of glass independently.

灯工玻璃工作室定时对外开放，在这里，你可以看到博物馆为将传统工艺植入现代实践而付出的不懈努力，专业人士和观众经过这一体验都会为传统工艺的历史感所折服。

The Flameworking Performance, in use at set times within the space, exemplifies the museum's efforts to incorporate traditional craft with contemporary practice, and sees experts and visitors engage in a modern take on a time-honoured technique.

博物馆概念设计的核心是营造一个充满乐趣的互动交流空间,为此专门整合了日间及晚间活动,这是特别针对宝山区这一较为偏远的选址而策划的。

从博物馆自行组织的展览到私人派对,各种附加活动共同构成一个真正意义上的动感空间。丰富多彩的体验选择不仅可以确保稳定的参观流量,还能吸引人们多次前来参与活动。而博物馆空间用作商业活动场所还将有助于场馆创收,保证其稳定的运营能力。

At the heart of the museum's concept design was the creation of a meaningful space for interaction and exchange. The integration of event spaces that offer the possibility for daytime and evening events was a key feature in this regard, particularly in the planning of a venue in Shanghai's relatively remote district of Baoshan.

From museum-organized temporary exhibitions through to private get-togethers, a programme of additional activities creates a 'living' space, not only ensuring a steady flow of visitors, but also giving them a reason to return. Moreover, the opportunity to use the building as an event location for commercial parties generates income and supports the museum's economic viability.

二层:
展示, 吸引, 赞赏

博物馆首层通过对特定空间的分隔来引导观众, 而二楼展厅则更加开放和宽敞。极简但不乏奢华的陈列柜展示着日益增多的本土和国际艺术家的玻璃藏品。步入此间的观众犹如置身画廊。一处特别开辟的空间为创作、采购或转借艺术品以及公众讲座等特色活动提供了场地, 其布局深意在于改变与成长。夹层透露出设计者的别具匠心与高瞻远瞩: 观众在这里可以直接俯瞰底层历史悠久的展品, 这种距离感正好强化了博物馆的核心策略——"讲述玻璃的历史、当下与未来"。尊贵专属的基调是为夹层后侧被称作"珍宝馆"的小型暗室而特意营造的。这一奢华空间可灵活用于举办小规模珍宝展。

LEVEL 2:
SHOWCASING, MESMERIZING, ADMIRING

Where the museum's first floor guides visitors through a specific grid of spaces, the second floor exhibition space is more open and expansive. A growing collection of glass art by contemporary local and foreign glass artists is displayed in showcases with a straightforward but luxurious design, creating a gallery-like atmosphere. The spatial layout suggests growth; there is room left for newly-created, purchased or loaned art objects, as well as for lectures and events. The physical design reinforces its forward-looking concept: visitors can literally look down to the first floor's historical exhibits, its distance underlining the museum's core strategy of telling the past, present and future of glass. At the back of the space, the so-called Jewelry Box, is a small, low-lit area where temporary exhibitions of smaller-sized treasures are held, creating a sense of exclusivity.

二楼陈列着专为展览量身订做的一系列艺术品，由此可见博物馆对其主旨的重视。一个"玻璃屋"位于通往二楼的走道当中，为Venini和Baccarat等世界顶级玻璃与水晶品牌打造了专门的展示空间。与此类顶级设计品牌建立联系也是博物馆长期发展的重要战略：不仅要将当代性与关联性植入藏品，还要有助于提升上海玻璃博物馆的国际形象。

这个通透"玻璃屋"的墙体由一千多个银色小瓶装点而成，每只小瓶都刻有一个汉字，它们共同讲述了一段古典爱情故事——"玻璃山"。此外，该空间的地板也为独家定制，由复杂的玻璃熔化工艺制作而成。"玻璃屋"也可用于举办小规模的私人派对，这里融合了当代艺术作品、古代口述历史及现代文化活动，是博物馆"衔接历史、当下与未来"的最佳体现。

The second floor is defined by an array of commissioned art pieces, tailor-made for the space and accentuating the museum's emphasis on storytelling. The House of Glass forms the gateway to the second floor and creates a showcase for unique pieces by amongst others Venini and Baccarat. Connecting with designer brands such as these is central to the museum's long-term strategy: not only does it inject contemporary context and relevance to the collection, it also helps raise the profile of Shanghai Museum of Glass on an international stage.

The walls of this transparent 'pavilion' are adorned by over one thousand custom-made silver-chrome bottles, each engraved with a single Chinese character to reveal the classic western love story, 'The Glass Mountain'. Also used to accommodate small-scale, exclusive events, the floor of the space is custom-made using a complex glass-fusing technique. In merging modern art, ancient oral history and present day events, The House of Glass exemplifies the museum's intention of connecting past, present and future.

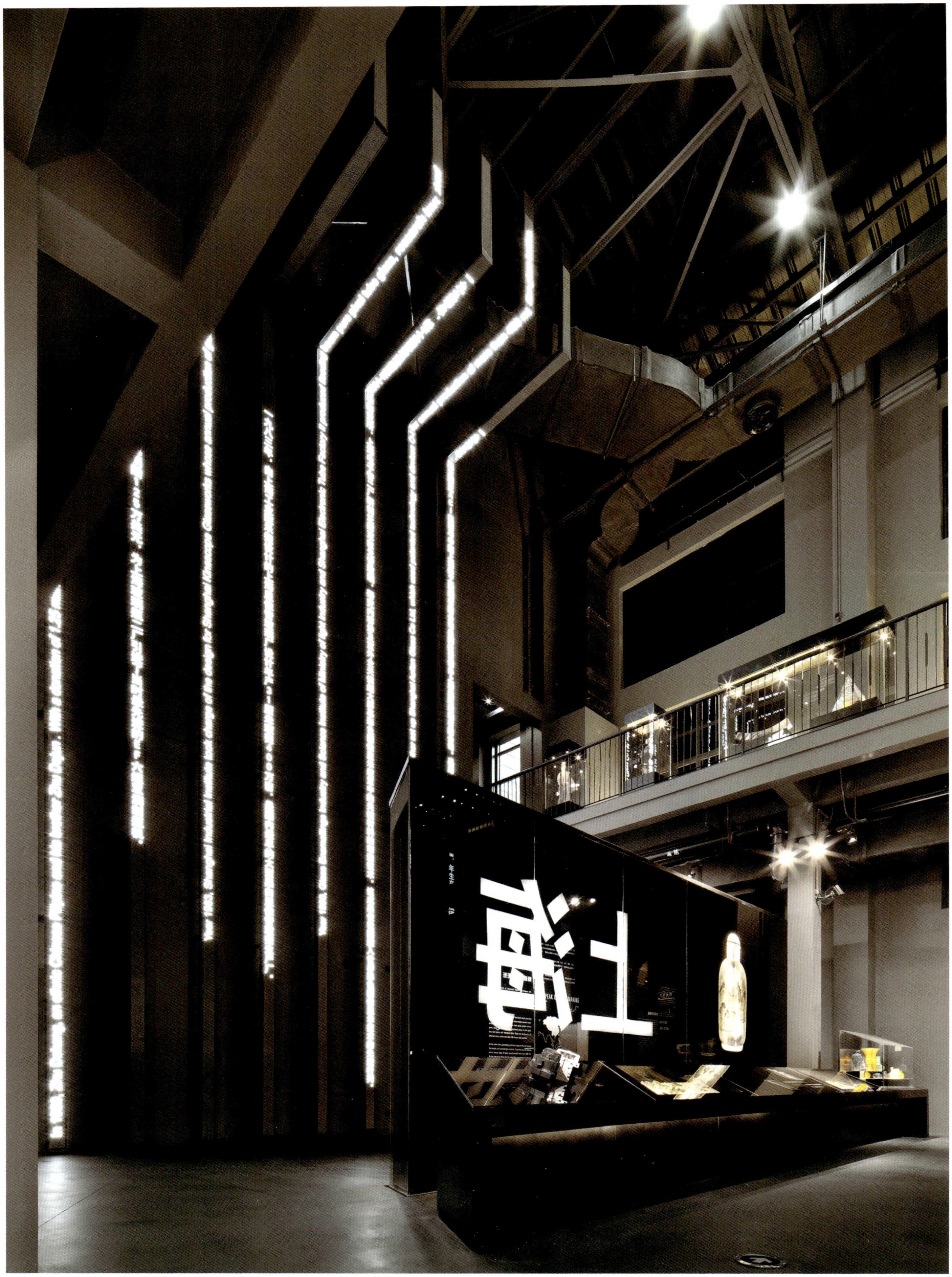

委托定制的艺术品以更具功能性的方式塑造着博物馆空间。博物馆一楼和二楼具有不同的外观和气质，但却通过与天花板齐高的LED艺术装置实现了自然衔接，灯光反射在两层的玻璃显示屏上，交相辉映。一个艺术装置由荧光灯管造型而成，呈现出契诃夫的名句——"不要告诉我月亮在闪耀，让我看到玻璃碎片上的光"，现场效果既惊艳又震撼。该作品不仅集艺术性与叙事性于一体，触及天花板的高度还突出了空间开阔性，提醒着观众正置身于原汁原味的玻璃老厂房建筑里。

Commissioned art pieces are additionally used in a more functional way to shape the museum space. The museum's first and second floor differ in look and feel but maintain a sense of continuity through a ceiling-height LED art installation whose lights are reflected in the glass display cases of both levels. A fluorescent text-based artwork depicting Anton Chekhov's famous line, 'Don't Tell Me The Moon is Shining; Show Me The Glint of Light on Broken Glass' combines artistry with storytelling to striking effect. Moreover, its positioning on the ceiling accentuates the height of the space, drawing attention to the original features of the old factory that have been deliberately left intact.

此类作品无不体现出上海玻璃博物馆通过与艺术家通力协作以支持艺术创作的重要策略。努力实践当中的艺术家驻馆计划，邀请国内外年轻艺术家到博物馆现场生活和工作，此举可谓对博物馆推动艺术发展策略的完美诠释。该计划显然是明智的共赢之举：可以有效地培育年轻与新晋玻璃艺术家；可为博物馆引入专门技艺和专业知识，在增强观众体验度的同时展示玻璃艺术的延续性与重要性；确保二楼艺术画廊和展览空间获得源源不断的艺术佳品，是丰富博物馆藏品的一种可持续手段。

Commissioned art pieces are central to Shanghai Museum of Glass' strategy of supporting the arts through collaborations and cooperation. Illustrating this concept is an ongoing artist-in-residence programme that invites young international talents to live and work on-site. The scheme presents several advantages: first and foremost, it is a way of nurturing fresh talents. Second, it imports specific skills and expertise to the museum, boosting visitor experience and demonstrating glass art's continued relevance. Finally, it ensures a constant stream of new artworks to the upper floor gallery and temporary exhibition space, offering a sustainable means of developing the museum's collection.

公共区域的作用

上海玻璃博物馆设置了多个可免票进入的公共区域，由此实现的开放性和活跃度充分展现了其社会服务性、包容性和亲和力。免费公共区域包括入口大厅和售票区、问询处、礼品店、连接博物馆两部分建筑的玻璃中庭以及二楼的咖啡厅。

除玻璃中庭与售票区之间的礼品店之外，所有公共区域均位于博物馆前方新建的L形建筑内，由玻璃中庭连接新老建筑。它们构筑成光线充足的自然空间，与博物馆展示厅幽暗的内部环境形成鲜明对比。这样的设计有利于营造出友善的环境，而且在参观人流较大时，这些区域还可成为绝佳的会集地或休憩区。

除提供舒适环境之外，公共区域还有两大附加功能：它别具匠心地衬托出所要展示的玻璃制品，激发参观者的好奇心并吸引他们入内；此外，公共区域活力四射而又极具魅力的氛围，在强化社区博物馆概念的同时，形成了当今中国文化机构中独一无二的空间形象。

THE ROLE OF PUBLIC AREAS

The Shanghai Museum of Glass incorporates several public areas that are accessible without purchasing a ticket. Their openness and energy underline fundamental elements of the venue's commitment to community, inclusiveness and accessibility. Specifically, these public spaces comprise the main lobby and ticketing area, an information desk, a shop, a glass void connecting the two parts of the museum building; and on the second floor, the museum café Murano@SHMOG.

With the exception of the museum's shop – located between the void and the ticketing area – all public areas are housed in the newest additions to the site's architecture, the L-shaped wrap-around building at the front of the museum and the glass atrium that links the two original structures. As such, they allow for bright, sunlit spaces that contrast starkly with the dimly-lit interiors of the museum main hall. This in itself creates a welcoming environment, with the areas acting as meeting or resting points in the wider visitor flow.

Comfort aside, they currently serve two additional purposes. Firstly, they incorporate carefully chosen features that allude to the glass exhibits on show within, driving visitors into the main museum hall. Equally important, they help create a lively atmosphere, reinforcing the concept of a community museum and setting the venue apart in China's cultural landscape.

入口大厅不仅具备卫生间和寄存柜等实用元素,还有舒适的沙发和座椅供参观者休憩,极具现代感和戏剧感的大吊灯也突出了博物馆的整体主题。简约宽大的票务台可根据参观流量和类型开启同步通道。首层还设有一个小型问询处,这是作为国家4A级博物馆必不可少的服务设施,强化了其地标性观光场所的显著地位。

The entrance area or lobby, for example, features not only practical elements such as bathrooms and storage lockers, but also comfortable sofas and seating for visitors, as well as a dramatic, contemporary chandelier that alludes to the theme of the entire museum. An extra wide ticketing desk facilitates visitor flow by accommodating simultaneous channels according to type. Also located on the ground floor is a information bureau that offers a practical and necessary addition for the museum's National AAAA Level Tourist Attraction status, reinforcing the venue's role of 'local landmark'.

为方便游览，入口区域正后方的博物馆商店环中央岛而建，还设有一个小型咖啡站。店内酒吧，毗邻连接博物馆两座建筑的玻璃的中庭，内有弹性座椅，参观者可以在那儿饮水休憩，这里的材质和设计都强化着博物馆基于玻璃这个极为现代的主题。

For ease of browsing, the museum shop that lies directly behind the entrance area is built around a central island, incorporating a small coffee-making station. A bar element within the shop and flexible seating inside the adjacent void – the glass atrium that links the two museum buildings – offers room to visitors to enjoy their drinks, its material and design reinforcing the museum's contemporary approach to its subject: glass.

游客无须购买门票便可以进入二楼咖啡厅消费。同L形新建筑的其他空间一样，咖啡厅布置得开放通透，不仅为游客提供了惬意的休闲空间，也成为了周边上班族钟爱的就餐地点。同样，中庭和大堂的内部设计均采用绚丽的玻璃元素，咖啡厅墙上还悬挂着"玻璃之城"穆拉诺 (Murano) 的迷人风景，这里使用的大型吊灯为意大利顶级玻璃品牌Venini。

上海玻璃博物馆公共区域不仅可以供游人休息，还带来了额外的收入。随着基础设施的改善，宝山办公租金和商业零售额预计将会增加，上班族将频繁光顾博物馆咖啡厅、商店和其他免费公共空间，他们最终将转化为消费人群，甚至惠及正在规划和开发中的GLASS+玻璃主题园。

Also accessible without ticket purchase is the second floor café Murano@SHMOG. Like other areas of the new, L-shaped building, its set-up is open and airy. Comfortable seating makes the café an attractive option to museum visitors, and a local alternative for nearby office workers. In the same way the void and lobby incorporate striking glass design into their respective interiors, the café does so through framed photographs of glass city Murano and a Venini chandelier.

Not only do the public spaces in the Shanghai Museum of Glass serve a practical purpose in providing areas for visitors to rest, they also offer an alternative source of revenue. It is hoped that as Baoshan's office and retail quotient increases along with improved infrastructure, workers will frequent Murano@SMHOG, shop and other free-of-charge spaces, ultimately converting to paying visitors or even future clientele of the wider GLASS+ Park planned developments.

设计理念的延伸: 开创GLASS+ 玻璃主题园

上海玻璃博物馆的塑造与设计不仅体现了博物馆核心理念, 更成就了博物馆的宗旨与目标: 在成功展示不断增加的藏品的同时, 还辅以细致入微的解读来丰富参观者的体验, 这在博物馆叙事方面体现得尤为明显。展望未来, 诸如珂庐 (KILN) 会所和Keep it Glassy国际创意玻璃设计展等策划项目的出现一定可以令这里成为高端活动的理想举办地, 从而催生新社会关系, 创造玻璃更多的可能性。

DESIGN CONCEPT ROLL-OUT: EXTENDING THE GLASS+ PARK

Thoughtfully created to reflect the museum's core concepts and support its key goals, the design of the Shanghai Museum of Glass not only successfully showcases the venue's growing collection, it also enriches the visitor experience by adding subtle layers of interpretation - particularly in the realm of storytelling. Moving forward, new initiatives such as KILN and Keep it Glassy support future activities in a space specifically designed to spark new relationships and possibilities related to glass.

在国际上的玻璃博物馆里风行着一种颇具特色的表演项目——热玻璃吹制表演。在这里，观众能参与到玻璃制作的过程以及玻璃吹制的艺术体验中，观察玻璃艺术家将又热又红、呈熔融状态的玻璃转变成晶莹剔透的艺术精品，深切地感受到这种工艺过程中的炽烈炉温。生动有趣的热玻璃吹制表演也是上海玻璃博物馆体验活动的重要环节，为博物馆打造多元化理念奠定了基石。热玻璃表演厅设在博物馆最古老的建筑之内，配置有玻璃制作的专业设备。

A shared characteristic of glass museums around the globe is the hot glass show. Here, visitors can take part in the process of glass making and the art of glass blowing, watch how glass artists turn red-hot molten glass into fragile artworks and experience first hand the intense heat involved in this craft. Its live and engaging shows, make the Hot Glass Performance of the Shanghai Museum of Glass an essential part of the museum experience, and a corner stone of the multifunctional museum concept. The Hot Glass Performance Hall is located in the oldest building on site and is equipped with professional glass making equipment.

DIY创意工坊为各年龄段的儿童与成人提供了不同的教程和工作坊。丰富多彩的活动课程包括玻璃彩绘、玻璃喷砂、玻璃马赛克等,侧重激发儿童及家庭参观者的参与热情。诸如热玻璃表演、DIY创意工坊等一系列特色项目的出现,明确了博物馆作为全天候游览目的地的品牌定位,同时完美呼应了博物馆致力于为观众提供趣味参观体验的宗旨。

The DIY Creative Workshop offers classes and workshops for children and adults of all ages. Courses include Glass Painting, Frosted Glass and Glass Mosaic and provide a means for the museum to engage particularly with young visitors and families, both identified as key target groups in the museum's longer-term strategy. Like the Hot Glass Performance, the DIY Creative Workshop reinforces the museum's positioning as a viable day out and underpins the museum's commitment to provide an educational experience for visitors.

本全部图片)
博物馆轴线上的创意工坊
Images (all):
Hands on workshops in the DIY Creative Workshop, located on the museum axis.

专为高端客户打造的私人会所——珂庐（KILN）位于博物馆一个老建筑内，竭力为有私密活动空间需求的博物馆VIP服务。作为博物馆社交及扩展人脉的场所，珂庐充分发挥了博物馆充满活力的社交功能。在这里，高贵古典的Venini照明与房门、天花板与定制壁图交相辉映。源自19世纪的玻璃技艺构建出了极具现代感的装饰图案，将历史与当下紧密相联。珂庐（KILN）时尚灵动的现代空间与博物馆主展览厅相得益彰，其间同样融入了兼具历史与现代气息的专属元素，同时还将世界上无数有关玻璃的故事向人们娓娓道来。

Designed with high-end users in mind, private members' club KILN is housed in a renovated building and targets visitors seeking an exclusive, private space for events. Used as a place to build relationships and forge connections, it underpins the community aspect of this living museum. Vintage Venini lighting is presented as a backdrop to custom-designed graphics on doors, ceilings and walls. Rearranged into modern decorative patterns, they incorporate elements from 19th century technical patent drawings of industrial and scientific glass. A contemporary space with sleek details, KILN's design creates an effect similar to the main museum hall, incorporating a sense of place; custom-made, historic and modern elements; and exploring myriad stories related to the world of glass.

Keep it Glassy国际创意玻璃设计展也采用了异曲同工的设计方法，其所处的建筑将在园区进一步开发时拆除，目标是吸引来自全球各地、阅历渐丰的年轻游客。该空间集合了砖墙表皮、混凝土地面和架空横梁，巧妙地保留了工业老厂房的原始样貌，从而暗含了该展览的潜在目的——当代设计师运用传统工艺探索玻璃的可能性。闪烁的镜面水池式展台营造出美不胜收的未来主义空间，其设计整体上比博物馆主厅更加前卫、轻灵，更适合举办第三方活动。

The same approach is taken in temporary exhibition space Keep it Glassy, housed in a building due for demolition ahead of the site's next phase of development and aimed at attracting a younger, cosmopolitan audience. The space's exposed brickwork, concrete floors and overhead beams serve as a reminder of the original function of the building, an industrial workshop. This connects to the underlying motive of the exhibition: exploring ways in which modern designers apply traditional processes of glassmaking. In contrast, sleek, mirrored pools serve as displays for exhibited objects, creating a modern and futuristic space. An altogether edgier alternative to the main museum hall, it is also suitable for hosting third-party activities and events thanks to an airy and flexible set-up.

博物馆品牌 MUSEUM BRANDING

品牌战略

中国正在经历一场来势汹汹的"博物馆风潮",仅在2011年就有386家主题各异的博物馆诞生。然而,若干城市的博物馆运营现状表明,这些新建的文化目的地并非个个门庭若市,有的人气完全无法与电影院、酒吧和购物中心相提并论。中国居民没有在周末闲暇时间参观博物馆的文化习惯,专家指出主要原因在于博物馆奉行的"精英主义"策略导致其曲高和寡,此外艺术教育的缺失也限制了观众的鉴赏能力。博物馆,特别是当代艺术博物馆,对中国普通居民来说,似乎是难以企及的文化制高点。

上海玻璃博物馆正在尝试扭转这一现象,因而高度重视亲和力与体验性这两大要素,对观众本身的知识储备也完全没有要求。极具针对性的品牌战略要求博物馆通过诸多现代手法吸引最广泛的受众。设计的现代感赋予了它极高的辨识度,令其得以从中国众多博物馆中脱颖而出。出人意料地在工业环境中培育文化热点是上海玻璃博物馆品牌战略的亮点,现代的博物馆建筑、创新的活动策划以及绚丽的现场视觉都是实现这一战略的有效途径。

玻璃博物馆的视觉设计经过精雕细琢,在呼应博物馆主题的同时又绝不与其他同时代博物馆雷同。凭借炫目的外观、现代的室内设计、风格统一的标识与宣传材料,上海玻璃博物馆一举成为宝山乃至上海市的地标性建筑,堪称中国文化机构中的楷模。博物馆惊艳的设计获得了国际重要刊物和诸多设计奖项的青睐,迅速提升了国际形象与影响力。许多中国观众用大量照片记录着各自的参观体验,并通过社交媒体发到网络上,更使得博物馆不拘一格与热情好客的特质深入人心。

由于博物馆地理位置较为偏远,品牌传播范围比市内同类型的博物馆更大,而选址宝山区也意味着博物馆不可避免地流失了一些习惯在市中心来去自由、步行观光的游客。但博物馆采取了诸多有效措施鼓励潜在观众专门抽出时间,使用代步工具前来宝山现场参观。如此一来,上海玻璃博物馆无疑会成为其他休闲服务提供商强大的竞争对手。博物馆必须做到让每位参观者坚信:选择这里便意味着最体贴的服务与最精彩的体验。

BRANDING STRATEGY

China is experiencing a so-called museum boom, with 2011 alone seeing the opening of 386 institutions spanning a range of themes. However, as numerous Chinese cities have discovered, the building and opening of new museums does not necessarily guarantee visitors, with such venues considered less attractive than cinemas, bars and shopping malls. The reasons behind Chinese visitors' reluctance to factor museum-going into their regular downtime is complex, but experts point to both museums' perceived elitism, and in the case of contemporary art venues in particular, an assumed inability to engage with works thanks to a lack of arts education in schools.

The Shanghai Museum of Glass seeks to counter this trend by placing a strong emphasis on accessibility and experience, regardless of visitors' prior knowledge of glass. The concept is conveyed by a targeted branding strategy that communicates the museum's modern approach to engaging a broad audience, and distinguishes the museum from its contemporaries in China. A key element of the branding strategy relates to cultivating the image of a cultural hotspot in an unexpected industrial setting. The message is supported both by the museum's modern architecture and design, its programme of activities, as well as by how the venue presents itself visually.

The visual identity of Shanghai Museum of Glass has been carefully crafted to reinforce its concepts and set the venue apart from its contemporaries. Through a striking façade, modern interiors and consistency in logos and promotional materials, the museum distinguishes itself through design, becoming a beacon for Baoshan district, a landmark for its city and a role model in China's cultural landscape. Its eye-catching design has earned the museum a global reputation through internationally significant publications and awards. Amongst Chinese audiences in particular, the museum's strong identity has resulted in numerous photographs documenting their visits being shared on social media networks, further reinforcing the venue's image of a non-conformist, welcoming space.

Because of the museum's relatively remote location, it arguably pushes its brand message to an even greater extent than comparable venues elsewhere in Shanghai. Its Baoshan setting means the Shanghai Museum of Glass misses the inevitable foot-traffic of its downtown counterparts, making spontaneous visits unlikely. Rather, potential audiences need to be encouraged to make time to travel to the museum's Baoshan location. Branding therefore forms an important tool to compete with other leisure providers, and serves to assure visitors that they will be looked-after, engaged and stimulated on an attractive, worthwhile day out.

服务理念

大约二十年前，包括上海轻工玻璃公司在内的上海工业企业开始搬迁至城郊，如今的上海玻璃博物馆正是借用了玻璃厂的旧址。这里原本是典型的工业制造场所，而今，服务门类的发展扩大提高了人们的工资待遇，促进了教育产业和创意产业的发展，同时还催生了中国新一代中产阶级。有鉴于此，对服务和生活方式等概念的投资是确保上海玻璃博物馆在日新月异的国际都市中得以宾客盈门的基础。

友好、专业和人性化是博物馆实现其主要功能的关键，它超越了简单的展品参观，而是致力于打造多元化的时尚生活中心。博物馆呈现了一系列与玻璃艺术有关的体验和活动，包括公众讲座、联谊活动、表演、电影放映、珂庐 (KILN) 私人会所和博物馆会员计划等。入口大厅的游客中心是博物馆拓展其社会关系的具体实践，亦是体现中国4A级博物馆高端品质的最佳窗口。

正在进行的各项计划展现了博物馆在创新背景下以人为本的理念，强大的运营团队为执行这些目标进行着不懈努力，以扬长避短。工作人员处于观众体验服务的第一线，资质优秀、训练有素的工作人员对场馆良好声誉的形成和维护起着至关重要的作用。工作人员的态度构成了博物馆的形象，并能最终决定首次参观者是否愿意再度前来。由于地处相对偏远的宝山区，前往博物馆最方便的交通工具是私家车，步行参观的可能性不大。因此，博物馆的服务水平必须超越上海其他同类竞争者。

GLASS+玻璃主题园在未来将推出全新的婚礼中心和商务园区，卓越的服务将为这个时尚生活中心吸引到越来越多的人气。博物馆面向追求卓越生活品质的受众进行品牌定位，从而成功吸引到构建园区所需的创意人士，这是园区走向未来，打造全新创意中心的第一步。

SERVICE CONCEPT

By virtue of its buildings, the Shanghai Museum of Glass is in part the result of industry starting to edge out of the city some twenty years ago – including that of glass. In place of manufacturing, service sectors have expanded to cause an increase in wages, prompting growth in areas such as education and the creative industries and creating a new Chinese middle class. With this in mind, investing in a service and lifestyle concept was fundamental to the Shanghai Museum of Glass' future security and relevance in a changing city.

A friendly, professional and 'human' face is key to a core function of the museum: fostering relationships. This is achieved through a lifestyle-centred approach that goes beyond a straightforward museum visit. Visitors are offered a broad experience and package of activities, all related to glass and art. A programme of events, for example, including lectures, networking events, performances and films screenings, as well as specific permanent initiatives such as private members' club KILN and a museum membership program have all been created specifically with service in mind. Extending the concept of relationship-building further still, a visitor centre housed in the museum's entrance area raises the venue to AAAA status in China's tourism rankings.

Whilst these ongoing schemes exemplify the museum's people-centric approach in the creative arena, just as important is a strong operational team to support the museum's goals and strengths. The first line in visitor experience, finding and training the right staff was fundamental to the venue's building and maintaining of a reputation of excellence. Integral to creating a positive visitor experience, their attitude shapes perceptions of the museum, and ultimately may determine whether or not individuals choose to return. Located in remote Baoshan, a destination most conveniently reached by private vehicle, spontaneous visits are unlikely, and for that reason, the museum's service element arguably needs to go above and beyond that of comparable institutions elsewhere in Shanghai.

Looking ahead, exemplary service will become increasingly important in attracting new audiences to the wider Glass+ Park as its additional lifestyle components are launched – the most important of which being phase two's wedding venue and business park. Branding the museum as a destination for an audience with a taste for quality and excellence is just a first step in building a creative community for the creative centre yet to be completed.

图注
CAPTIONS

照片来源
PHOTO CREDITS

罗昂
LOGON
12, 16, 17~19, 28, 29, 32, 34, 35, 37, 52, 62, 64

上海轻工玻璃有限公司
SHANGHAI GLASS COMPANY LTD
20, 21

DIE PHOTODESIGNER 摄影
DIE PHOTODESIGNER
38, 39, 41, 46, 50, 58, 60, 66, 68, 70, 82, 83, 86, 88, 90 left, 92, 93, 95, 100, 102, 104, 105~108, 109 left bottom,
109 right, 110, 114 top, 122, 124, 126, 127~130, 131, 132, 134, 136, 137 left & centre, 138, 140, 141, 142, 144,
145 right, 148, 152, 153, 154, 156, 158, 159

夏宇
CHARLIE XIA
52, 75, 76, 78, 80, 81, 84, 85, 87, 90 right top and left, 91, 98, 109 top left, 112, 113, 114 bottom, 116, 117, 118,
119, 120, 137 right, 145 left top & bottom, 146, 149, 150, 162, 164

SETH POWERS 摄影
SETH POWERS PHOTOGRAPHY
91 bottom

协调（亚洲）
COORDINATION ASIA LTD
14, 15, 30~31, 115

项目信息
PROJECT CREDITS

城市规划，建筑设计：罗昂建筑设计
Urban Planning, Architectural Design: logon urban. architecture. design
www.logon-architecture.com

艺术指导，博物馆设计，视觉辨识设计：协调（亚洲）
Art Direction, Museum planning, VI Design: COORDINATION ASIA Ltd.
www.coordination-asia.com

承包方：上海美术设计有限公司
Contractor: Shanghai Art-Designing Co., Ltd.

业主：上海轻工玻璃有限公司
Client: Shanghai Glass Co., Ltd.

地址：中国上海长江西路685号
Location: 685 West Changjiang Rd Shanghai / China
www.shmog.org